Phase Transitions in Cell Biology

Gerald H. Pollack · Wei-Chun Chin
Editors

Phase Transitions in Cell Biology

Springer

Editors
Gerald H. Pollack
Department of Bioengineering
Box 355061
University of Washington
Seattle WA 98195
USA

Wei-Chun Chin
PO Box 2039
Merced CA 95344
USA

ISBN 978-1-4020-8650-2 ISBN 978-1-4020-8651-9 (eBook)

Library of Congress Control Number: 2008928026

Printed on acid-free paper

9 8 7 6 5 4 3 2 1

springer.com

Preface

Why phase transitions?

Phase transitions occur throughout nature. The most familiar example is the one that occurs in water: the abrupt, discontinuous transition from a liquid to a gas or a solid, induced by a subtle environmental change. Practically magical, the ever-so-slight shift of temperature or pressure can induce an astonishing transition from one entity to another that bears little resemblance to the first.

In a sense, this transition is a kind of amplification: a subtle change inducing a radical change. So "convenient" a feature is seen throughout the domains of physics and chemistry, and one is therefore led to wonder whether it might also be common to biology. Capitalizing on this kind of amplification would seem to be a sensible approach for Mother Nature to have taken as she built the domain of life. Small change yields big response. What could be more sensible?

The idea that such a phenomenon might be pervasive throughout biology was first proposed in a book written by one of us (GP). *Cells, Gels and the Engines of Life* proposed that the phase transition is indeed a central protagonist in the drama of life. Many of the most fundamental cellular processes are arguably attributable to radical structural shifts triggered by subtle changes that pass above a critical threshold. These processes include transport, motion, signaling, division, and other fundamental aspects of cellular function.

Largely on the basis of that book, a symposium was organized in Poitiers, France, to bring together people who have additional evidence for the role of phase transitions in biology, and this book is largely, albeit not completely, a compendium of some of the more far-reaching presentations. Contributions from several scientists who were unable to attend are included as well.

The book should be suitable for anyone interested in the nature of biological function, particularly those who tire of lumbering along well trodden pathways of pursuit, and are eager to hear something fresh. The book is replete with fresh interpretations of familiar phenomena, and should serve as an excellent gateway to deeper understanding.

It is our hope to awaken to the reader the idea that phase transitions play a role as important in biology as they play in physics and chemistry. Biology, after all, is little more than applied physics and chemistry. Is it not?

Seattle, WA, USA
Merced, CA, USA

Gerald H. Pollack
Wei-Chun Chin

Contents

On the Reversible Abrupt Structural Changes in Nerve Fibers Underlying Their Excitation and Conduction Processes 1
Ichiji Tasaki

Nonequilibrium Phase Transition in Scattered Cell Communities Coupled by Auto/Paracrine-Like Signalling 23
H. Berry

Interfacial Water Compartments on Tendon/Collagen and in Cells 43
I.L. Cameron and G.D. Fullerton

The Role of Ion-Exchange on Trypsin Premature Activation in Zymogen Granules 51
Y.X. Ding, E. Chen, K. Yang and W.-C. Chin

Whole-Cell Phase Transition in Neurons and its Possible Role in Apoptotic Cell Death 63
F. Gallyas and J. Pál

Puzzles of Cell and Animal Physiology in View of the Chain-Ordering Transition in Lipid Membrane 73
D.P. Kharakoz

Ephemeral Gels: The Biological Example Applied to a New Type of Polymers 95
J. Picard, S. Giraudier and V. Larreta-Garde

The Cytoskeleton of the Living Cell as an Out-of-Equilibrium System 111
Guillaume Lenormand, Adriano M. Alencar, Xavier Trepat, En-hua Zhou, Ben Fabry, James P. Butler and Jeffrey J. Fredberg

Unexpected Linkage Between Unstirred Layers, Exclusion Zones, and Water 143
Gerald H. Pollack and James Clegg

"Autothixotropy" of Water – An Unknown Physical Phenomenon, and its Possible Importance for the Cytoskeleton 153
Bohumil Vybíral and Pavel Voráček

Propagation of Volume Phase Transitions as a Possible Mechanism for Movement in Biological Systems 159
L. Yeghiazarian and R. Lux

Cell Plasma Membranes and Phase Transitions 171
Mark M. Banaszak Holl

Index 183

Contributors

Adriano M. Alencar
Molecular and Integrative Physiological Sciences, Department of Environmental Health, Harvard School of Public Health, Boston, MA, USA

Mark M. Banaszak Holl
Department of Chemistry and Program in Macromolecular Science and Engineering, Michigan Nanotechnology Institute for Medicine and Biological Science, 930 N. University Avenue, University of Michigan, Ann Arbor, MI 48109-1055

H. Berry
INRIA, Team Alchemy, Parc Club Orsay Université, 3, rue J. Rostand, 91893 Orsay Cedex France, hugues.berry@inria.fr

James P. Butler
Molecular and Integrative Physiological Sciences, Department of Environmental Health, Harvard School of Public Health, Boston, MA, USA

I.L. Cameron
University of Texas Health Science Center, San Antonio, USA, CAMERON@uthscsa.edu

E. Chen
School of Engineering, University of California, Merced, CA, 95344, USA

W.-C. Chin
School of Engineering, University of California, UC Merced, P. O. Box 2039, Merced, CA, 95344, USA, wchin2@ucmerced.edu

James Clegg
Section of Molecular and Cellular Biology, and Bodega Marine Laboratory, University of California, Davis, 2099 Westside Road, Bodega Bay 94923, USA

Y.X. Ding
Department of Bioengineering, University of Utah, Salt Lake City, UT 84112, USA

Ben Fabry
Molecular and Integrative Physiological Sciences, Department of Environmental Health, Harvard School of Public Health, Boston, MA, USA

Jeffrey J. Fredberg
Molecular and Integrative Physiological Sciences, Department of Environmental Health, Harvard School of Public Health, Boston, MA, USA

G.D. Fullerton
University of Texas Health Science Center, San Antonio, USA

F. Gallyas
Department of Neurosurgery, University of Pécs, H-7623 Pécs, Rét utca 2, Hungary

S. Giraudier
Errmece, University of Cergy Pontoise, BP 222, 95302 Pontoise, France

D.P. Kharakoz
Institute of Theoretical and Experimental Biophysics, Russian Academy of Sciences, Institutskaya street 3, 142290 Pushchino, Russia

V. Larreta-Garde
Errmece, University of Cergy Pontoise, BP 222, 95302 Pontoise, France

Guillaume Lenormand
Molecular and Integrative Physiological Sciences, Department of Environmental Health, Harvard School of Public Health, Boston, MA, USA, glenorma@hsph.harvard.edu

R. Lux
Department of Dentistry, UCLA, Los Angeles, CA, USA

J. Pál
Department of Neurosurgery, University of Pécs, H-7623 Pécs, Rét utca 2, Hungary

J. Picard
Errmece, University of Cergy Pontoise, BP 222, 95302 Pontoise, France

Gerald H. Pollack
Department of Bioengineering, Box 355061, University of Washington, Seattle WA, 98195, USA, ghp@u.washington.edu

Ichiji Tasaki
Bldg. 13, Rm. 3E-25, LIMB, NICHD and LCMR, NIMH, National Institutes of Health, Bethesda, MD, 20892, USA, itasaki@erols.com

Xavier Trepat
Molecular and Integrative Physiological Sciences, Department of Environmental Health, Harvard School of Public Health, Boston, MA, USA

Pavel Voráek
Lund Observatory, Box 43, SE-221 00 Lund, Sweden, postmaster@astro.lu.se

Bohumil Vybíral
Department of Physics, Pedagogical Faculty, University of Hradec Králové, Rokitanského 62, CZ-500 03 Hradec Králové, Czech Republic, Bohumil.Vybiral@uhk.cz

K. Yang
Department of Physiology, University of Toronto, Toronto, Ontario, M5S 1A8, Canada

L. Yeghiazarian
Department of Bioengineering, currently at the Department of Biostatistics, School of Public Health, UCLA, Los Angeles, CA, USA

En-hua Zhou
Molecular and Integrative Physiological Sciences, Department of Environmental Health, Harvard School of Public Health, Boston, MA, USA

On the Reversible Abrupt Structural Changes in Nerve Fibers Underlying Their Excitation and Conduction Processes

Ichiji Tasaki

Abstract The cortical gel layer of nerve fibers has the properties of a cation-exchanger. Hence, this layer can, and actually does, undergo a reversible abrupt structural change when monovalent cations (e.g. Na^+) are substituted for the divalent counter-ions (e.g. Ca^{2+}). This structural change brings about a sudden rise in the water content of the layer which in turn produces a large enhancement of cation mobilities accompanied by a shift of ion-selectivity in favor of hydrophilic cations. Based on these grounds, it is argued that the electrophysiological processes known as "nerve excitation and conduction" are, basically, manifestations of abrupt structural changes in the cortical gel layer. In recent studies, we have shown that several aspects of the excitation phenomena can actually be reproduced by using synthetic polyanionic hydrogels in place of living nervous tissues. It is noted that these studies of synthetic model systems lead us to a better understanding of the process of divalent-monovalent cation-exchange in natural and artificial polyanionic gels.

Keywords Nerve excitation and conduction · structural phase transition in nerve fiber · divalent-monovalent cation-exchange

1 Introduction

The polypeptide chains in solutions can be reversibly converted, as is well known, from the random coil to the helical form. Hydrogen bonds formed between different groups in one long polypeptide chain lead the whole chain into the helical form. This structural transformation is very sharp; that is, a change of a few degrees in temperature or a few percent of solvent composition is sufficient to complete the transformation. Hence, the term "phase transition" has been employed to describe this reversible structural change (see Doty and Yang, 1956; Zimm and Bragg, 1959; Ptitsyn et al., 1968).

I. Tasaki
National Institutes of Health, Bethesda, MD, 20892, USA
e-mail: itasaki@erols.com

G.H. Pollack, W.-C. Chin (eds.), *Phase Transitions in Cell Biology*,

Negatively charged polyelectrolyte gels in salt solutions can be converted from the swollen state to the compact state when the monovalent counter-ions are replaced with divalent cations. This structural transformation is also sharp; that is, it can be initiated and completed by a small change in the salt composition of the surrounding solution (see Katchalsky and Zwick, 1955; Kuhn, 1962; Tanaka, 1981; Tasaki and Byrne, 1992). There seems little doubt that such reversible, abrupt changes in the gel structure are a process of very common occurrence in various biological systems.

The objective of the present article is to demonstrate that the reversible abrupt structural changes occurring in the cortical gel layer of nerve fibers are at the base of the process of excitation and conduction. There is abundant evidence to show that these structural changes are associated with divalent-monovalent cation-exchanges taking place in the negatively charged sites in the cortical gel layer.

2 Abrupt Structural Changes in Synthetic Polyanionic gels

In this section, attempts are made at elucidating the role of divalent-monovalent cation-exchange in the production of abrupt structural changes in polyanionic gels. To achieve this end, we describe, in some detail, the results of several observations which we have made on synthetic polyanionic gels during recent years. In the following section, we treat the results obtained from living nerve fibers on the basis of our knowledge about the abrupt structural changes in synthetic gels.

2.1 Discontinuous Volume Transition

We now know that the volume of a small piece of cross-linked Na-polyacrylate or Na-polymethacrylate gel can change *discontinuously* when Na-ions in the gel are replaced, by gradual steps, with divalent cations (Tasaki and Byrne, 1992, 1994; Tasaki, 1999). In most of our studies, we have examined the effects of application of Na- and Ca-salts to the gel. However, we have observed on several occasions that similar results can be obtained by using other divalent cations, such as, Mg^{2+}, Sr^{2+}or Ba^{2+} in combination with other monovalent cations, such as, Li^{+}, K^{+}, Rb^{+}, Cs^{+} or tetraalkylammonium ions.

An example of our results is shown in Fig. 1. Here, small spherical beads of cross-linked polyacrylate gel of approximately the same diameter were placed in a series of petri dishes containing 40 mM NaCl solution (kept at pH 7.4). Small aliquot quantities of a concentrated $CaCl_2$ solution was added to the dishes, and the diameters of the gel beads were determined when equilibrium was reached between the gel beads and the surrounding salt solution.

It is seen in the figure that the Ca-salt added to the solution produced a gradual decrease in the gel diameter initially. However, at the point where the Ca^{2+} concentration in the dish rose to about 1.2 mM, there was a *discontinuous decrease* in the

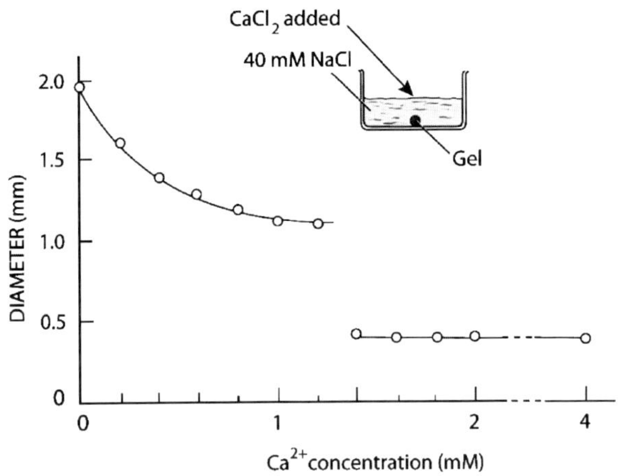

Fig. 1 The diameter of a small spherical gel bead immersed in a 40 mM NaCl solution, plotted against the concentration of Ca^{2+} added to the solution (Tasaki and Byrne, 1992)

gel diameter. The volume of the gel bead fell at this point by a factor of roughly 1/10. We now wish to know how this discontinuous fall in the gel volume is brought about.

As the first step toward achieving our goal, we made measurements of the quantities of $Na^+$and Ca^{2+} *inside* individual gel beads, as a function of the concentration of the Ca-salt in the dish. After some initial difficulties, we were able to obtain reasonably reproducible results, indicating that the quantity of Na^+ *inside the bead* falls and that of Ca^{2+} rises smoothly (but rather precipitously) as the Ca^{2+} concentration *outside the beads* is raised gradually from zero. As expected from the existence of a high density of fixed negative charges in the gel beads, the sum of the quantities of $Na^+$and Ca^{2+}(expressed in unit of equivalents) was found to remain constant within the experimental uncertainty. We *could not* detect any sign of discontinuity in the quantity of Na^+ or Ca^{2+} *inside the gel beads* in the entire range of Ca^{2+} concentration outside.

It is known that the COO^--groups in macromolecules overwhelmingly prefer Ca^{2+} to Na^+ (Williams, 1970; Levine and Williams, 1982). In fact, we have seen that, at the point of discontinuity of the gel volume, roughly 80% of the entire negative charge *inside the gel bead* was neutralized by $Ca^{2+}$and only about 20% by Na^+. We have noted already that, at this point, the external concentration of Ca^{2+} was only about 1/33 (=1.2/40) of that of Na^+. Such a remarkably high selectivity for Ca^{2+} had never been considered in previous studies of living excitable tissues.

We expect that the external Ca^{2+}-concentration required for inducing a discontinuous volume transition varies according to the NaCl concentration in the dishes. In fact, when the NaCl concentration in the dish was raised from 25 to 150 mM, there was a nearly proportionate rise in the required external Ca^{2+} concentration. Consequently, the concentration ratio $[Ca^{2+}]/[Na^+]$ at the point of volume discontinuity was found to be insensitive to the variation in the NaCl concentration in the dish.

Interestingly, this ratio $[Ca^{2+}]/[Na^{+}]$ at the point of volume discontinuity appears to vary inversely with the density (and probably with the regularity of distribution) of the fixed negative changes in the gel. We have seen that lowering the pH of the surrounding salt solution down to 5 or less brings about a noticeable increase of this ratio. Furthermore, copolymerization of acrylic acid with acrylamide in gel synthesis was found to bring about a marked rise of the ratio $[Ca^{2+}]/[Na^{+}]$, even when the uncharged component (acrylamide) was only 10% of that of the negatively charged component (acrylic acid). No macroscopic discontinuity was observed in the gel volume when the acrylamide content was 33% or more.

Finally, we note in Fig. 1 that, in the range of Ca^{2+} concentration *higher than* about 1.3 mM, the size of the gel bead was practically independent of the external Ca^{2+} concentration. It is known that Ca-ion is capable of forming a complex, bridging $C00^{-}$-group of one chain with another COO^{-}-group in a neighboring chain (see Williams, 1970). The size of such a complex is probably determined roughly by the exclusion volumes of the hydrocarbon chains, the Ca-ions, the ligands and water molecules involved. Thus, we visualize the compact, Ca^{2+}-rich structure of the gel bead as stabilized by Ca-bridges between the COO^{-}-groups in the gel bead (see Tasaki, 2005a).

2.2 Propagation of the Boundary Between Swollen and Compact Regions of a Gel Strand

We now describe a phenomenon that is directly related to the *instability* of the structure which constitutes the boundary between the compact and swollen regions of a gel strand. When one end of a strand of cross-linked polyacrylate gel in its compact (Ca^{2+}-rich) state is immersed in a NaCl solution, swelling of the gel starts at this end, thus creating a gel strand which consists of two structurally distinct regions, compact and swollen. The transitional region, or the boundary between the two regions, of the gel strand was found to be remarkably short and sharp. By using such a gel strand, it was found possible to induce continuous displacement of the boundary with the aid of an electric current applied to the strand.

The diagram at the top of Fig. 2 shows the arrangement employed. Here, a gel strand in its compact state was placed across a 10 mm wide platform separating the solution of NaCl from the $CaCl_2$ solution in a plastic chamber. After covering the surface of the gel strand with a thin layer of liquid paraffin, an electric current was delivered to the strand. The current was directed from the portion of the gel strand immersed in the NaCl solution toward the other end immersed in the $CaCl_2$ solution. As expected, the portion of the gel in the NaCl solution began to swell. And then, the boundary between the compact and swollen portions started to move towards the compact side.

With this experimental setup, there was a layer of salt solution on the surface of the gel strand and also on the platform. Consequently, the current employed was considerably stronger than the intensity expected from a simple $2Na^{+} \rightarrow Ca^{2+}$

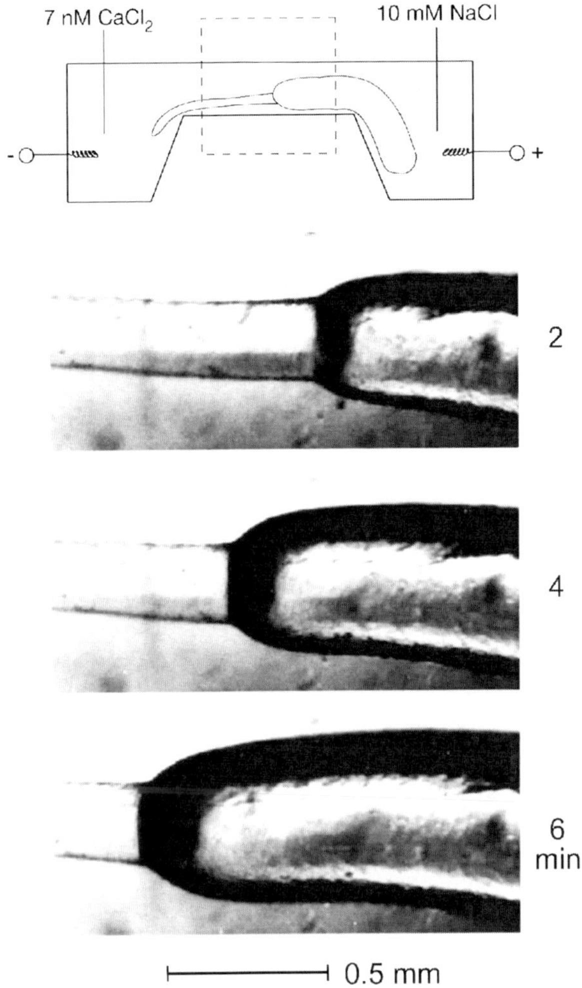

Fig. 2 Photomicrographs showing displacements of the boundary between the compact and swollen regions of a gel strand induced by electric current of 1 mA (Tasaki, 2002)

exchange at the sharp boundary. When the boundary was located on the platform, propagation of the boundary from the compact side of the strand to the swollen side could be induced by application of a current flowing in the reverse direction.

2.3 Formation of Bundles of Polyelectrolyte Chains

We now describe an interesting pattern of binding of divalent cations to the negatively charged chains in polyelectrolyte gels. We first demonstrate that exposure of a swollen polyelectrolyte gel to a solution of divalent cation salt can produce highly refractile bundles of polymer chains in the gel (Tasaki, 2005b).

Strands of cross-linked Na-polyacrylate gel in their swollen state are wholly transparent and there is no structure recognizable in the strand under dark-field illumination. Individual polymer chains in the gel are highly hydrated and a considerable portion of the Na-ions in the gel are loosely associated with the chains (see Kern, 1939; Huizenga et al., 1950; Ikegami, 1964). When a swollen gel strand is exposed to a $CaCl_2$ solution, the surface of the gel immediately becomes visible, and soon highly refractile bundles of polymer chains begin to appear on the surface. These bundles originate usually from the cut end or some irregular spots of the gel surface and spread gradually into the interior of the gel stand.

The photomicrograph in Fig. 3, left, shows highly visible bundles of polymer chains stretching from the compact region to the swollen region of the gel strand. Because of a large change in the diameter of the strand at the boundary, the polymer chains near the surface are under tension which tends to align the chains in the longitudinal direction. This parallel alignment of the polymer chains is considered as the condition favorable for creating Ca^{2+}-bridges and forming bundles of polyelectrolyte chains.

When Ca-ions are delivered directly into the interior of a swollen gel by the aid of a glass pipette, a quite different pattern of bundle formation is observed (see Fig. 3, right). Here, a pipette filled with a $CaCl_2$ solution was pushed into the gel and an outwardly directed current was delivered to the gel by using an Ag-AgCl wire inserted in the pipette. The photomicrograph of the bundle formation in the figure was taken shortly after delivering a 0.1 mA current for about 15 s. It is noted in the figure that the pattern of bundles reflects the distribution of the stretched polymer chains created by insertion of the pipette. It is noted also that the process of bundle formation initiated by the Ca-ions in the vicinity of the orifice of the pipette spreads *cooperatively* along these chains.

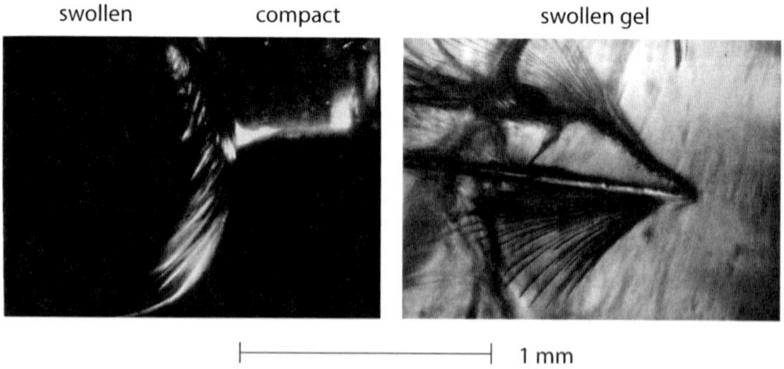

Fig. 3 *Left*: Highly refractile bundles of polymer chains formed in the transitional zone between the compact and swollen regions of a cross-linked polyacrylate gel strand. *Right*: Bundles of polyelectrolyte chains formed inside a swollen gel by application of an electric current by use of a glass pipette containing a $CaCl_2$ solution (Tasaki, 2005b)

Since these bundles of polymer chains formed by the delivery of a current pulse are surrounded by many Na^+-acrylate units of the chains in the gel, they tend to fade away little by little within a few minutes. When, however, a pulse of an inwardly directed current is delivered to the gel, the portions of the bundles located in the vicinity of the orifice of the pipette promptly disappear as a consequence of the rapidity of the Ca^{2+}-Na^+ exchange process.

The formation of bundles of biopolyelectrolyte chains with polyvalent cations appears to be a quite general phenomenon in biology (see Tang et al., 1996).

2.4 Abrupt Changes in Electric Impedance of the Gel Associated with Ca^{2+}-Na^+ Exchange

When a compact gel layer undergoes a transition to a swollen state in association with a divalent-monovalent cation-exchange, there is a sudden change in the electric properties of the layer. The ion mobilities in the gel layer are greatly enhanced by the rise in the water content of the layer. It is therefore easy to demonstrate that the a.c impedance abruptly falls when a Ca^{2+}-Na^+ exchange induces a structural change in the superficial layers of a compact gel layer (see Fig. 4).

The diagram at the top of the figure schematically illustrates the setup employed. A 1 kHz a.c. was applied to a cylindrical gel rod by use of a pair of platinized platinum electrodes, placed one inside and the other outside the gel. The surface of the internal platinum electrode was completely insulated except for a short portion located inside the gel rod. The intensity of the a.c. was adjusted to give rise to an alternating voltage of about 20 mV (rms) across the gel layer. This voltage was amplified, half-wave rectified and was passed through a resistor-capacitor circuit. The resulting non-alternating (d.c.) voltage output was taken as a measure of the impedance of the layer.

It is seen in the record shown in the figure that, following application of a 100 mM NaF solution to the surface of a compact gel rod, there were repetitive abrupt falls of the impedance. Note that *fluoride* (or *phosphate*) salts, instead of *chloride* salts, of monovalent cations had to be employed in these experiments. It is known that, in the *lyotropic series* of common anions, F^- and $HPO_4{}^{2-}$ are the anions most effective in precipitating various proteins (see Tasaki et al., 1965). We visualize the abrupt structural change observed in polyanionic gels as a transition from a *compact* state to a *swollen* (i.e. *hydrated*) state. Reflecting its position in the lyotropic series, chloride salts tend to raise the water content of the gel; consequently, it is unfavorable to evoke abrupt changes in the a.c. impedance of the gel repetitively.

Abrupt changes in the a.c. impedance can be demonstrated in a compact gel sheet (about 1 mm in thickness) compressed between two thin plastic plates, each of which has a small (about 1 mm in diameter) hole at a matching position. When one of the small surfaces of the layer is exposed to a 66 mM $CaCl_2$ solution and the other small surface to a 100 mM NaF solution, repetitive abrupt falls in the impedance followed by fairly *rapid recovery* are observed. Under these conditions,

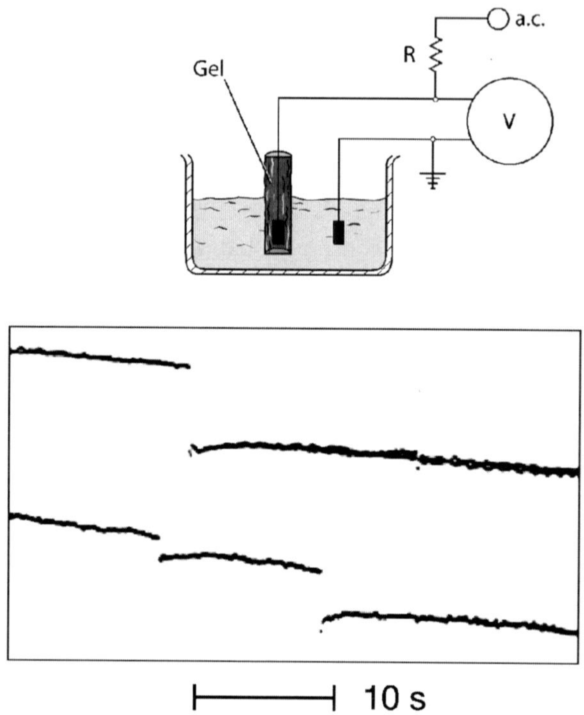

Fig. 4 Abrupt fall of the electric impedance across the superficial layers of a compact gel rod immersed in a 100 mM NaF solution. The impedance at 1 kHz was 1.1 kΩ, and the change observed in the upper trace was about 1% initially. (Tasaki, 2005b)

the gel surfaces exposed to the salt solutions are subjected to a considerable mechanical stress. This stress constrains the polymer chains near the gel surface and this is considered as the predominant factor that brings about a rapid recovery of the impedance loss.

2.5 Electric Potential Changes Associated with Abrupt Structural Transitions

The variation of the potential difference across a compact gel layer can be induced by application of a monovalent cation salt under a variety of experimental conditions (Tasaki, 2005b). An example of those observations is presented in Fig. 5. Here a compact, Ca-rich gel rod was compressed by means of two plastic plates separating a NaF solution from a $CaCl_2$ solution. The plastic plate facing the NaF solution had a small (0.5 mm diameter) hole, and the other plate had of a larger hole at the matching position.

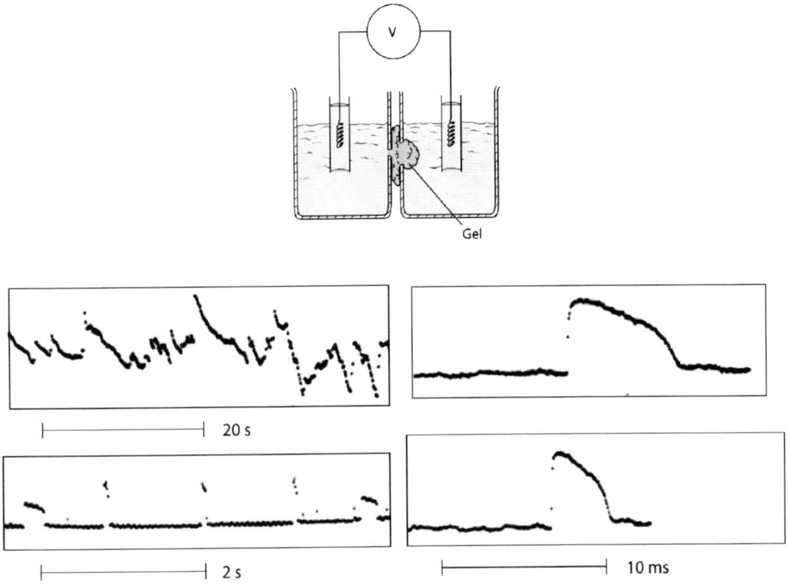

Fig. 5 Variation of the potential difference across a compressed gel layer that were exposed to a 100 mM NaF solution on one side and to a 66 mM $CaCl_2$ solution on the other side. The amplitudes of the abrupt potential changes were between 1 and 24 mV (Tasaki, 2005b)

Many records of the potential variation taken under these experimental conditions were found to have time-courses that resemble those taken from living nerve fibers (see the figure). A sharp rising phase, followed by a slower falling phase, is characteristic of these "responses" of the compact gels. The variability of the amplitude of these recorded "responses" is considered to arise from the non-uniformity of the size of the activated patches on the gel surface.

It is possible to evoke similar "responses" by application of an electric current to a thin layer of polyacrylate gel separating salt solutions containing monovalent and divalent cations. To record such "responses", however, A.C.-coupling of the output signals from the gel layer to the recording preamplifier input is required, because the applied current generates a large, slowly varying potential drop across the layer.

3 Excitation Processes in Nerve Fibers

This section is devoted to the description of several observations demonstrating that the electrophysiological phenomena known as "nerve excitation and conduction" are, basically, manifestations of reversible abrupt structural changes occurring in the cortical gel layer of the nerve fiber. The process of divalent-monovalent cation-exchange in the layer assumes the principal part in the present discussion.

3.1 The Role of Ca^{2+} in Nerve Excitation: Jacques Loeb's Theory

In 1883, Sidney Ringer published his recipe of the saline solution which is favorable for maintaining the excitability of excised heart muscles and pointed out the importance of Ca-salt in the solution. In the year 1900, Jacques Loeb published an article subtitled "The poisonous character of a pure NaCl solution" and showed that the "poisonous" effect of a pure NaCl solution on excitable tissues can be counteracted, or "antagonized", by addition of the salt of Ca-ions.

Loeb formulated at that time a quite comprehensible theory of nerve excitation. It is highly instructive to see how Loeb visualized the process of nerve excitation a century ago. The following sentences are taken from his paper and monograph: "The salts, or electrolytes in general, do not exist in living tissues as such exclusively, but are partly in combination with proteins or fatty acid. The salts or electrolytes do not enter into this combination as a whole, but through ions. The great importance of these ion-proteid compounds (or soaps) lies in the fact that, by the substitution of one ion for another, the physical properties of the proteid compound change (p. 327 in Loeb, 1900)". "The normal irritability of animal tissues depends upon the presence in these tissues of Na, K, Mg and Ca ions in the right proportion; – any sudden change in the relative proportions – gives rise to an activity or an inhibition of activity (p. 95 in Loeb, 1906)". "The quotient of the concentration of Na ion over the Ca ions, C_{Na}/C_{Ca}, becomes therefore of importance for phenomena of irritability (p. 79 in Loeb, 1906)".

From these sentences, we can see that Loeb had a fairly good grasp of the process of exchange of Ca^{2+} for Na^+ or K^+ in living excitable tissues. It may appear very strange that his theory did not gain much popularity among classical neurophysiologists. The reason for this absence of wide acceptance of his theory might be, in part at least, that the experimental evidence cited in support of his idea was very indirect in nature and probably was not quite compelling. However, the main reason appeared to be that most physiologists at that time were passionately preoccupied with W. Nernst's mathematical theory of nerve excitation dealing with an abstract semipermeable membrane (see Nernst, 1908). Consequently, Loeb's theory of nerve excitation remained dormant for a long time to come.

3.2 The Process of Ca^{2+}-Na^+ Exchange at the Base of Action Potential Production

We have seen already that cross-linked polyanionic hydrogels can generate abrupt potential changes in response to a Ca^{2+}-Na^+ exchange. We now demonstrate that squid giant axons are capable of producing action potentials in association with a simple Ca^{2+}-Na^+ exchange occurring in the cortical gel layer of the axons (see Tasaki, 1982; 1999).

Here, we choose our observations made on squid giant axons under intracellular perfusion. The diagram at the top of Fig. 6 schematically illustrates the experimental

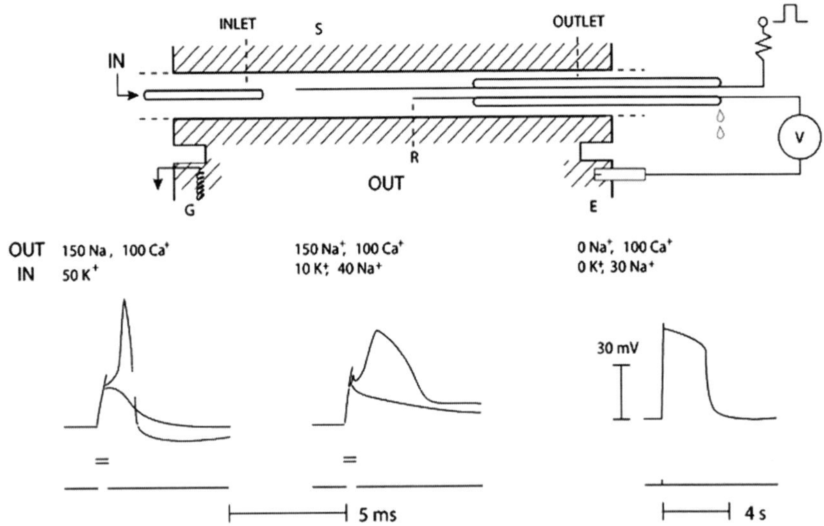

Fig. 6 *Top*: Schematic diagram illustrating the setup for intracellular perfusion of a squid giant axon. *Bottom*: Action potentials recorded from an axon. The concentrations of the cations in the extracellular (OUT) and intracellular (IN) salt solutions are indicated (Tasaki, 1982)

arrangement employed. A giant axon (about 0.5 mm in diameter) is mounted on a plastic platform (30 mm wide). The major portion of the protoplasm inside the axon between the tips of the two cannulae ("inlet" and "outlet" in the diagram) has been removed beforehand (by suction). The inlet cannula is connected to a reservoir of the solution for perfusion (containing KF, NaF, glycerol, etc). The flow of the solution inside the axon (roughly 25 μl/min) is maintained by adjusting the height of the reservoir. The external surface of the axon is exposed to a Ca^{2+}-containg salt solution. Pulses of outwardly directed current are delivered across the cortical layer of the axon by use of an internal metal wire electrode (S in the diagram) and a ground electrode (G). The responses of the layer to the current pulses are recorded with non-polarizable electrodes (R and E).

The left-hand record in the figure was taken from an axon which was intracellularly perfused with a K^+-salt solution and immersed in a chloride salt solution of Ca^{2+} and Na^+. We see that a brief pulse of outward current evoked an action potential which is very similar to those observed in intact axons (without intracellular perfusion). Under these conditions, we can maintain the ability of the axon to generate full-sized action potentials for more than 10 h.

Shortly after the invention of the technique of intracellular perfusion, we found that the excitability can be maintained in axons intracellularly perfused with solutions of the fluoride or phosphate salts of Cs^+, Na^+, choline, tetraalkylammonium ions, etc, as long as the pH and the osmolarity of the intracellular solution are kept in a proper range. Furthermore, it was found possible to evoke action potentials in axons immersed in the solution of the salt of Sr^{2+} or Ba^{2+} substituting Ca^{2+}. [Note, however, that the salt of divalent cations introduced into the axon interior irreversibly damages the axon.] *No action potential can be elicited from axons when the Ca-ions (or their substitutes) in the external solution are completely replaced with monovalent cations.*

The record furnished in Fig. 6, right, is an example the action potentials taken from axons intracellularly perfused with a dilute NaF solution and immersed in a $CaCl_2$ solution. Here, the cortical layer of the axon is sandwiched between a Na^+-salt solution inside and a Ca^{2+}-salt solutions outside. *There is no Na^+ outside the axon.* The concentration of Ca^{2+} in the external medium is high enough to maintain the cortical layer in the Ca^{2+}-rich, compact state. Hence, delivery of a pulse of outward current to the layer is expected to induce a $2Na^+ \rightarrow Ca^{2+}$ exchange in the cortical layer.

From the finding that all-or-none action potentials can be observed under these simplified experimental conditions, we conclude that the process of Na^+-Ca^{2+} exchange occurring in its cortical layer of the axon is responsible for the generation of the observed action potentials.

3.3 A Wave of Reversible Abrupt Structural Change Running Along the Cortical Gel layer

In 1970, in his Tilden Lecture entitled "The biochemistry of sodium, potassium, magnesium, and calcium", R. J. P. Williams predicted the existence of a mechanical change in the nerve fiber in association with a propagating nerve impulse. He said: "This (propagating nerve impulse) is often pictured as a physical event – electrostatic field changes altering the membrane so that it changes from a potassium to a sodium permeable condition. Could it not rather be that inward diffusion and binding of calcium causing a running wave of structural change along the membrane?" (p. 362 in Williams, 1970). It was known at that time that, during repetitive stimulation of a squid giant axon, radioactively labeled Ca-ions in the surrounding medium are rapidly incorporated into the axon (Hodgkin and Keynes, 1957). Later on, using the method of intracellular perfusion, it was shown that, at the peak of excitation, the influx of Ca^{2+}across the cortical gel layer rises to an intensity 200–300 times as high as its influx at rest (see p. 224 in Tasaki, 1982).

Ten years after William's prediction, we actually found an unmistakable sign of a running wave of structural change—*shrinkage* of the squid giant axon following the production of an action potential (Iwasa and Tasaki, 1980). At the same time, we were gratified to find, in addition, a definite sign of *swelling* of the axon which *preceded* the shrinkage. This discovery of *swelling* had great impact on our discussion of the process of nerve excitation.

The schematic diagram at the top of Fig. 7 illustrates the design of the piezoelectric sensor device employed for detection of small and rapid mechanical changes in squid giant axons. As is well-known, these axons are capable of responding to repetitive brief shocks with production of full-sized action potentials for a long period of time. The record shown below was taken after averaging a large number

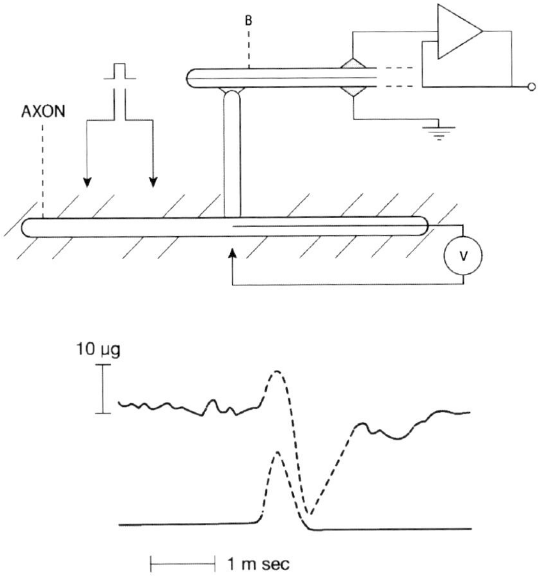

Fig. 7 *Top*: Schematic diagram illustrating the piezoelectric device for detecting small and rapid pressure changes associated with excitation of a squid giant axon. *Bottom*: Mechanical response and action potential recorded simultaneously (Tasaki 1982)

of responses. It is seen that the *swelling* of the axon starts almost simultaneously with the onset of the action potential. Furthermore, the peak of the swelling roughly coincides with the peak of the action potential.

This finding, obtained by use of a piezoelectric pressure sensor, was extended by employing a Fotonic sensor for detection of a small displacement of the axon surface (Tasaki and Iwasa, 1982). It was shown that, during the rising phase of the action potential, the axon surface moves outwards and this surface displacement is generally in the range between 0.5 and 2 nm.

From these observations, we deduce the following conclusion: *The propagating nerve impulse is a running wave of reversible structural change, representing a continuous displacement of the boundary between the site of swelling and the site of shrinkage of the cortical gel layer.*

The abrupt and drastic fall of the a.c impedance of the cortical layer associated with action potential production (Cole and Curtis, 1939) can now be safely attributed to the swelling (i.e. rise of water content) of the cortical layer running along the axon.

3.4 Spectral Analysis of Changes in Extrinsic Fluorescence Associated with Nerve Excitation

In 1968 we made the discovery that nerve fibers stained with appropriate fluorescent molecules can generate "optical responses" representing transient changes in

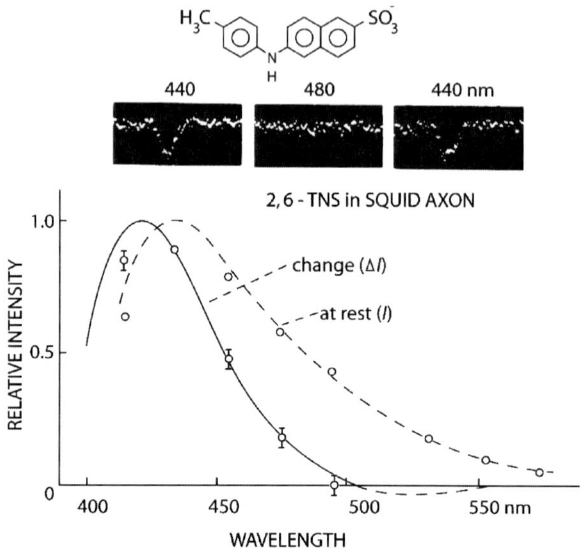

Fig. 8 *Top*: Samples of records of 2,6-TNS fluorescence changes associated with action potential production. *Bottom*: the wavelength dependence of fluorescence of 2,6-TNS injected into a squid giant axon (I) and that of fluorescence changes associated with action potential production (ΔI) (Tasaki et al., 1973)

intensity of the fluorescent light associated with action potential production (Tasaki et al., 1968). Our analyses of such optical responses have yielded much information about the structural changes which take place in the cortical layer of the nerve fiber during excitation (see Tasaki, 1982). In this subsection, we describe one of those analyses (see Fig. 8).

Anilinonaphthalene sulfonate (ANS) and toluidinylnaphthalene sulfonate (TNS) are known to emit intense fluorescent light when they are dissolved in organic solvent (with low solvent polarity), such as ethanol, but they do not fluoresce when dissolved in water. When water is added to an ethanol solution of these compounds, a fall in the fluorescence intensity is accompanied by a red-shift of the emission spectrum. Taking advantage of this property, ANS and TNS have been used for optically probing the *hydrophobicity* of the micro-environment of these molecules inside proteins (see Weber and Laurence; 1954; McClure and Edelman, 1966).

Figure 8 shows the results obtained by using 2,6-TNS injected into a squid giant axon. The broken line in the figure, marked (I), represents the spectrum of the fluorescent light emitted by the 2,6-TNS probe under constant illumination with polarized quasi-monochromatic light of 365 nm. The continuous line, marked (ΔI), shows the transient fall in the fluorescent light intensity brought about by electric stimulation of the axon, plotted against the wavelength of the emitted light. Note that the spectrum of ΔI is shifted to the shorter wavelength side of the spectrum I.

The significance of this finding can easily be understood on the basis of the Ca^{2+}-Na^{+} exchange process taking place in the cortical layer of the axon. In the resting (Ca^{2+}-rich) state of the axon, the micro-environment of the 2,6-TNS molecules in the layer must be *effectively hydrophobic,* because the fluorescence emission is intense. The transition of the cortical layer to its swollen state drastically enhances the *hydrophilicity* of the micro-environment of these probe molecules and brings about an abrupt fall in light intensity accompanied by a red-shift of the emission spectrum. We found that the observed spectrum of ΔI can be reproduced by subtracting the emission spectrum of this probe molecule dissolved in ethanol-water mixture from that observed in pure ethanol.

In this connection, it is to be remembered that the *hydrophobic* sites in the resting axon are most likely to take up K-ions far more easily than Na-ions (see pp. 343, 350 and 357 in Williams, 1970). In the squid giant axon at rest, addition of K-salt to the external medium readily causes a loss of excitability and a depression of the resting potential of the axon. We may now attribute this distinguishing property of K-ions to the *hydrophobicity* of the cortical layer of an axon at rest revealed by the observation described above.

In the swollen, *hydrophilic* state of the cortical layer, the cation mobilities are high and the selectivity for hydrophilic cations is enhanced. The behavior of Na-ions – and other monovalent cations such as hydrazinium (see p. 211 in Tasaki, 1982) – in the excited state of the axon may now be safely attributed to the running wave of swelling of the cortical gel layer.

3.5 Production of All-or-none Action Potentials by Electric Stimulation of Myelinated Nerve Fibers

In myelinated nerve fibers, the axis-cylinder of the fiber is covered, as is well known, by a layer of myelin sheath except at the nodes of Ranvier. The occurrence of the transient physiochemical events underlying the process of excitation and conduction is limited to the naked portion of the axis-cylinder at individual nodes of Ranvier (see Tasaki, 1982). The naked surface of the axis-cylinder at the node of a bullfrog motor nerve fiber is 0.5–1 μm wide and roughly 10 μm in diameter. Here, we analyze how action potentials are generated by electric stimulation of the node.

The records of action potentials developed by a *single node of Ranvier* furnished in Fig. 9 were taken by use of the following arrangements: The node under study (N_1 in the diagram) was kept in physiological saline solution (containing Na^+ and Ca^{2+}) and was separated from the proximal portion of the fiber (including N_0) by a narrow air-gap. The Ranvier nodes in this portion were rendered inexcitable by use of a dilute anesthetizing solution. The distal portion of the fiber (between N_1 and N_2) was used as a part of the "pipette electrode" for recording action potentials. Brief voltage pulses were applied between node N_1 and the proximal portion of the fiber, whereby generating pulses of outwardly directed current through the node under study, N_1.

It is seen in the figure that action potentials evoked were about 110 mV in ampli tude. It is noteworthy that the amplitude was practically independent of the duration

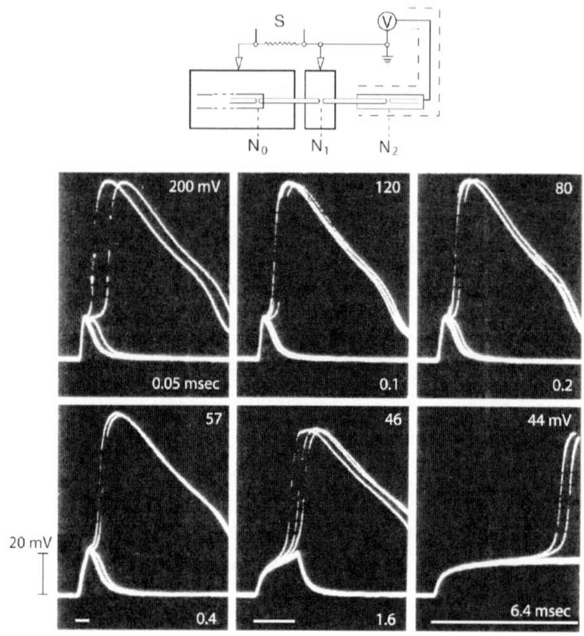

Fig. 9 Action potentials of a single node of Ranvier evokeded by application of brief voltage pulses. The duration and voltage of the stimulating pulse employed are indicated in each of the frames of record. Several records of the responses to the same stimulus strength are superimposed in each frame (Tasaki, 1982)

of the stimulating pulses, as long as the evoked action potential started after the end of the pulse. That is, *the response of a node of Ranvier to a brief stimulating pulse is all-or-none.* The threshold potential was about 22 mV above the resting potential and was also independent of the duration of the brief stimulating pulse employed.

It is known that the production of an action potential at the node is accompanied by a drastic fall in the electric resistance of the cortical layer at the node (see p. 69 in Tasaki, 1982). Hence, it is reasonable to assume that the cortical layer at the node is *swollen* and becomes *hydrophilic* in its excited state. In the resting state, the node is sensitive to "depolarization" by K-salt, implying that the layer is *compact* and effectively *hydrophobic*. Thus, the mechanism of excitation of a node of Ranvier is considered to be basically the same as that of other excitable tissues.

The potential fall from the peak of the action potential is a reflection of the "relaxation process" which is considered to arise primarily from accumulation of the extracellular cations in the intracellular space (see p. 68 ibid.). The *shoulder* of the action potential is regarded as representing the onset of a *transition* of the cortical layer from its swollen state (modified by the relaxation process) to the compact state (see p. 274 ibid.).

We now analyze the potential changes brought about *directly* by a brief stimulating pulse. It is noted that, at the end of the stimulating pulse of the threshold intensity, the observed potential change is close to the well-known value *RT/F* ($\approx$ 25 mV), which is not large enough to appreciably alter the cation distribution in a cation exchanger membrane (see p. 264 ibid.). By such a stimulating pulse, only a small fraction of the entire cortical layer at the node is expected to be thrown into the excited (i.e. swollen) state. The state of the node containing swollen spots or patches is unstable. As a consequence of the electric interaction of the swollen spots and patches with the remaining surface area of the node, the cortical layer tends to reach, eventually, either a uniformly swollen state or a uniformly compact state. Note the *bifurcation* of the potential trace in the records following the termination of the applied brief stimulating pulse.

We represent the fraction of the cortical layer at the node excited directly by the stimulating pulse by α. Then, the fraction of the remaining surface at the node area is $(1 - \alpha)$. The intensity of the inwardly directed current through the swollen portion of the surface is given by $\alpha(E_e - V)/r_e$, where E_e represents the e.m.f. of the layer in its excited state, r_e the resistance of the (entire) layer of the node in the excited state and V is the intracellular potential in the "mixed" state. Analogously, the outward current flowing through the remaining (resting) area is given by $(1-\alpha)(E_r - V)/r_r$, where E_r and r_r are the e.m.f. and the resistance of the layer in the compact (i.e. resting) state.

Immediately after the ends of the stimulating pulse of the threshold intensity, namely, during the period in which the *bifurcation* of the potential trace is observed, the net current through the node should vanish. It follows from this that

$$\alpha \frac{(E_e - V)}{r_e} + (1 - \alpha) \frac{(E_r - V)}{r_r} = 0 \tag{1}$$

By solving this equation for α, we obtain

$$\alpha = \frac{1}{\left(\dfrac{E_e - E_r}{V - E_r} - 1\right)\dfrac{r_r}{r_e} + 1} \tag{2}$$

In the present case, the action potential amplitude, $(E_e - E_r)$, is 110 mV, the potential rise evoked by threshold stimulation, $(V - E_r)$, is about 22 mV, and the ratio of the electric resistance of the layer at rest to that at the peak of excitation, r_r/r_e, is about 10. From these values, we find that the fraction α is about 1/40. The conclusion reached from this analysis is that *activation of a very small portion of the nodal area by an applied stimulus does* "trigger" *transition of the entire nodal surface to its excited state in an all-or-none manner.*

We know that a decrease in the Ca^{2+}-concentration in the external medium or some other chemical treatment of nerve fibers can cause lowering of the threshold stimulus strength. Hence, the fraction α can become even smaller than the value estimated above. We now have good understanding of the origin of the all-or-none behavior in the process of action potential production.

3.6 *Local Current, Cable Property and Conduction Velocity of Nerve Fibers*

In the Handbook of Physiology published in the year 1879, Ludimar Hermann advanced his *local current* (Strömchen) theory. He proposed in this book that nerve conduction is brought about by successive excitation of the neighboring resting area of the nerve by the local electric current generated by the excited area.

In the following years, the importance of this concept of re-stimulation by virtue of the local current has been fully recognized by every investigator studying the process of nerve conduction. However, the identity of the structural elements in the nerve (trunk) involved in the process of re-stimulation, "core conductor" and "sheath", remained completely obscure for a long time.

Shortly before and after World War II, the cable properties of both myelinated and non-myelinated nerve fibers were thoroughly clarified (see below), and the details of the process of re-stimulation by the local current became abundantly clear. In this subsection, we briefly describe how the velocity of nerve conduction is determined by the cable properties of the nerve fiber involved in the process of re-stimulation.

The electric resistance and capacitance of the myelin sheath were determined by recording the action current (and the quantity of electricity) passing through a short portion of an internodal (myelinated) segment under the condition that *two nerve impulses,* evoked near the two ends of the same fiber, *collide and vanish* at the site of the internode under study (see p. 72, Tasaki, 1982). The resistance of the portion of the myelin sheath was evaluated from the quantity of electricity which traverses the sheath during one complete cycle of rise and fall of the potential associated with the action potentials simultaneously generated at the two nearest nodes. (Note that the capacitive contribution to the measured quantity vanishes in this case.) The capacitance of the portion was determined from the potential and the quantity of electricity observed at the peak of the action potential, after subtraction of the resistive contribution at this moment. By applying the same method to a short portion of a nerve fiber including one node (rendered inexcitable by light anesthetization), the resistance and capacitance of the node at rest were estimated.

These measurements have shown that the resistance and capacitance of the myelin sheath of the amphibian motor nerve fiber are, respectively, $2.9 \times 10^7\ \Omega\,\text{cm}$ and 1.6×10^{-11} F/cm, the resistance-capacitance product being about 0.46 ms. This finding indicated that the myelin sheath does not behave like a good insulator when the potential difference across the sheath is varying rapidly with time. Because of the capacitive flow of electric current through the myelin sheath, the local current generated by abrupt excitation of one node does not reach the neighboring node instantly. Evidently, the major portion of the internodal conduction time (about 0.1 ms across a $2 \sim 2.5$ mm long internode) represents the time required for raising the potential inside the neighboring node to the level high enough to trigger a transition to its excited state (see p. 95 in Tasaki, 1982; Tasaki and Matsumoto, 2002).

We now turn to the cable property of squid giant axons. Owing to their great size (about 0.5 mm in diameter), the electric parameters of these axons can be determined more-or-less directly. The capacitance of the cortical layer (C) is shown to be about 1.0 $\mu F/cm^2$, the ohmic resistance of the layer at the peak of excitation (R^*) is $25 \sim 40\,\Omega\,cm^2$, and the resistivity of the axon interior (ρ) is $30 \sim 70\,\Omega\,cm$ (see Cole and Hodgkin, 1939; Hodgkin and Huxley, 1952).

We have seen that, when a brief stimulating shock is delivered to a squid giant axon, a traveling wave of rapid swelling is induced in the cortical layer of the axon. In the vicinity of the sharp boundary between the swollen (excited) and compact (resting) regions of the axon, the local currents are generated, spreading exponentially to the two sides of the boundary. The conduction velocity is determined by the spread of the local current in the vicinity of the boundary.

Some time after the inception of the technique of intracellular perfusion, Matsumoto and Tasaki (1977) examined the distribution of the local current in the squid giant axon in great detail and found that there is a simple quantitative relationship between the conduction velocity and the electric parameters. The equation describing the velocity, v, of an axon of diameter d is as follows:

$$v = \frac{1}{C}\sqrt{\frac{d}{8\rho R^*}}. \tag{3}$$

By introducing the above-stated values of the electric parameters, the velocity, calculated for axons of 0.05 cm in diameter, is found to be between 15 and 28 m/s. with an average of 22 m/s. The agreement between the calculated and observed values is good. It is noted that, in non-myelinated nerve fibers, the conduction velocity is directly related to the symmetric spread of the local current to the compact and swollen sides of the boundary between the two structurally distinct regions (see Tasaki, 2006).

4 Conclusion

(1) COO^- groups of polyelectrolyte chains in salt solutions bind Ca^{2+} preferentially and reversibly. The compact structure of a Ca^{2+}-rich polyelectrolyte gel is stabilized by calcium-bridges formed between polyelectrolyte chains.
(2) The cortical gel layer of nerve fibers or cells can be converted from the compact state to the swollen state by substituting monovalent cations (e.g. Na^+) for the divalent counter-cations (Ca^{2+}). The cortical layer is effectively hydrophobic in its resting state, and hydrophilic in its excited state.
(3) The process of action potential production is nothing but an electrical manipulation of reversible abrupt structural changes occurring in the cortical gel layer of nerve fibers and cells.

Acknowledgments The author expresses his gratitude to Dr. Peter Basser and Dr. Ralph Nossal of the Laboratory of Integrative and Medical Biophysics, NICHD, for their continuous support.

References

Cole, K. S. and Curtis, H. J., 1939. Electric impedance of the squid giant axon during activity. *J. Gen. Physiol.* 22, 649–670

Cole, K. S. and Hodgkin, A. L, 1939. Membrane and protoplasm resistance in the squid giant axon. *J. Gen. Physiol.* 22, 671–687

Doty, P. and Yang, J. T., 1956. Polypeptides. VII. Poly-γ-benzyl-L-glutamate: The helix-coil transition in solution. *J. Am. Chem. Soc.* 78, 498–500

Hermann, L., 1879. Allgemeine Nervenphysiologie in *Handbuch der Physiologie,* 1ster Theil, 1–196. F. C. W. Vogel, Leipzig

Hodgkin, A. L. and Huxley, A. F., 1952. A quantitative description of membrane current and its application to conduction and excitation in nerve. *J. Physiol. (London)* 117, 500–544

Hodgkin, A. L. and Keynes, R. D., 1957. Movement of labelled calcium in squid giant axons. *J. Physiol. (London)* 138, 253–281

Huizenga, J. R., Grieger, P. F., and Wall, F. T., 1950. Electrolytic properties of aqueous solutions of polyacrylic acid and sodium hydroxide. I. Transference experiments using radioactive sodium. *J. Am. Chem. Soc.* 72, 2636–4232

Ikegami, A., 1964. Hydration and ion binding of polyelectrolytes. *J. Polymer Sci.* A. 2, 907–921

Iwasa, K. and Tasaki, I., 1980. Mechanical changes in squid giant axons associated with production of action potentials. *Biochem.Biophys. Res. Commun.* 95, 1328–1331

Katchalsky, A. and Zwick, M., 1955. Mechanochemistry and ion exchange. *J. Polymer Sci.* 16, 221–234

Kern, W., 1939. Der osmotische Druck wässeriger Lösungen polyvalenter Säuren und ihrer Salze. *Z. phys. Chem.* A 184, 197–210

Kuhn, W., 1962. Ändeung von chemischen Gleichgewichten und Lösligkeitgleichgewichten bei mechanischer Dehnung von Gelen. *Koloid Z. u. Z. f. Polym.* 182, 40–50

Levine, B. A. and Williams, R. J. P., 1982. The chemistry of calcium ion and its biological relevance. In: *The role of calcium in biological systems* (L. J. Anghileri and A. M. Tuffet-Anghileri eds), CRC Press, Inc. Florida. pp. 3–26

Loeb. J., 1900. On ion-proteid compounds and their role in the mechanics of the life phenomena. I. The poisonous character of a pure NaCl solution. *Am. J. Physiol.* 3, 327–338

Loeb, J., 1906. *The Dynamics of the Living Matter*, Columbia University Press., New York

Matsumoto, G. and Tasaki, I. (1977) A study of conduction velocity in nonmyelinated nerve fiber. *Biophys. J.* 20, 1–13

McClure, W. O. and Edelman, G. M., 1966. Fluorescent probes for conformational states of proteins. I. Mechanism of 2-p-toluidinylnaphthalene-6-sulfonate, a hydrophobic probe. *Biochemistry.* 5, 1908–1918

Nernst, W., 1908. Zur Theorie des elektrischen Reizes. *Pflügers Arch. f. d. ges. Physiol.* 122, 275–314

Ptitsyn, O. B., Kron, A. K., and Eizner, Yu. Ye. 1968. The models of the denaturation of globular proteins. I. Theory of globula-coil transitions in macromolecules. *J. Polymer Sci.* C. 16, 3509–3517

Ringer, S., 1883. A further contribution regarding the influence of the different constituents of the blood on the contraction of the heart. *J. Physiol. (London)* 4, 29–42

Tanaka, T., 1981. Gels. *Sci. Am.* 244, 110–123

Tang, J. X., Wong, S., Tran, P. T., and Janmey, P. A., 1996. Counterion induced bundle formation of rodlike polyelectrolytes. *Ber. Bunsenges. Phys. Chem.* 100, 796–806

Tasaki, I., 1982. *Physiology and Electrochemistry of Nerve Fibers.* Academic Press New York

Tasaki, I., 1999. Rapid structural changes in nerve fibers and cells associated with their excitation processes. *Jpn. J. Physiol.* 49, 125–136

Tasaki, I. 2002. Spread of discrete structural changes in synthetic polyanionic gels: A model of propagation of a nerve impulse. *J. Theor. Biol.* 218, 497–505

Tasaki, I., 2005a. Abrupt structural changes in polyanionic gels evoked by Na-Ca ion exchange: Their biological implications. *Macromol. Symp.* 227, 97–104

Tasaki, I., 2005b. Repetitive abrupt structural changes in polyanionic gels: A comparison with analogous processes in nerve fibers. *J. Theor. Biol.* 236, 2–11

Tasaki, I., 2006. A note on the local current associated with the rising phase of a propagating impulse in nonmyelinated nerve fibers. *Bull. Math. Biol.* 68, 483–490

Tasaki, I. and Byrne P. M., 1992. Discontinuous volume transition in ionic gels and their possible involvement in the nerve excitation process. *Biopolymers.* 32. 1019–1023

Tasaki, I. and Byrne, P. M., 1994. Discontinuous volume transition induced by calcium-sodium ion exchange in anionic gels and their neurobiological implications. *Biopolymers.* 34, 209–215

Tasaki, I., Carbone, E., Sisco, K., and Singer, I., 1973. Analyses of extrinsic fluorescence of the nerve membrane labeled with aminonaphthalene derivatives. *Biochim. Biophys. Acta.* 323, 220–233

Tasaki, I. and Iwasa, K., 1982. Rapid pressure changes and surface displacements in the squid giant axons associated with production of action potentials. *Jpn. J. Physiol.* 32, 69–81

Tasaki, I. and Matsumoto, G., 2002. On the cable theory of nerve conduction. *Bull. Math, Biol.* 64, 1069–1082

Tasaki, I., Singer, I., and Takenaka, T., 1965. Effects of internal and external ionic environment on excitability of squid giant axon. A macromolecular approach. *J. Gen. Physiol.* 48, 1095–1123

Tasaki, I., Watanabe, A., Sandlin, R., and Carnay, L, 1968. Changes in fluorescence, turbidity and birefringence associated with nerve excitation. *Proc. Nat. Acad. Sci. U.S.A.* 61, 883–888

Weber, G. and Laurence, D. J. R., 1954. Fluorescent indicators of adsorption in aqueous solution and on the solid phase. *Biochem. J.* 56, 31

Williams, R.J. P., 1970. Tilden Lecture. The biochemistry of sodium, potassium, magnesium and calcium. *Quart. Rev. Chem. Soc.* 24, 331–365

Zimm, B. H. and Bragg, J. K., 1959. Theory of the phase transition between helix and random coil in polypeptide chains. *J. Chem Phys.* 31, 526–535

Nonequilibrium Phase Transition in Scattered Cell Communities Coupled by Auto/Paracrine-Like Signalling

H. Berry

Abstract Auto/paracrine cell-to-cell communications via diffusive messengers can be coupled to a positive feedback loop in which cell stimulation by a messenger results in the production of new messengers. This yields a potential mechanism for relay transmission of the emitted message. This paper investigates the influence of noise on this mutual coupling of the cells with their environment, using numerical simulations of a stochastic minimal model. The results demonstrate that the deterministic (mean-field) approximation of this stochastic process fails short of predicting its behaviour because of the presence of strong noise-induced fluctuations. Instead, the behaviour of the model can be explained by the occurrence of a nonequilibrium phase transition, which is found to be in the universality class of directed percolation. This provides a theoretical framework to understand signal transmission in these stochastic systems.

Keywords Signal transmission · autocrine relay · stochastic models · critical phenomena · directed percolation

1 Introduction

Complex behaviours in cell communities such as self-organization and emergent phenomena may result from coupling of the cells with an environment they dynamically modify. For instance, cells often respond to molecules in their environment via intracellular signalling pathways that eventually result in altered concentration of the very extracellular molecular species that triggered the pathway.

A well-known example is auto/paracrine signalling. In this paradigm, cells emit a peptidic factor (e.g. EGF) that diffuses in the extracellular space until it reaches a neighbouring cell (paracrine signaling) or the source cell that emitted it (autocrine signaling) (Wiley et al., 2003). Stimulation by the diffusive factor may in turn trigger

H. Berry
INRIA, Team Alchemy, Parc Club Orsay Université, 3, rue J. Rostand, 91893 Orsay Cedex France
e-mail: hugues.berry@inria.fr

G.H. Pollack, W.-C. Chin (eds.), *Phase Transitions in Cell Biology*,

intracellular signalling cascades (e.g. the MAPK pathway) that eventually lead to the release of new diffusive factor molecules in the environment (positive feedback loop) (Pribyl et al., 2003a; Freeman, 2000).

Strikingly, this process can be considered as a rather generic communication pattern, that is not restricted to members of the EGF family or cytokines, but is also encountered in cAMP-mediated communications between *Dictyostelium discoideum* cells; acylated homoserine lactone-mediated quorum sensing in *Vibrio fischeri* (James et al., 2000) and other bacteria (de Kievit & Iglewski, 2000); or even airborne virus spreading – which has been implied in the dissemination of the foot-and-mouth disease (Gloster et al., 2005) or the influenza virus flu (Hammond et al., 1989) for instance.

Although dissimilar, these examples share common features. First, the messenger travels at random, through diffusional or nondirectional transport. Moreover, encounter with the messenger molecule alters cell functioning in such a way that new messengers are ultimately excreted. Finally, stimulation by the messenger molecules is usually followed by a refractory or lag period during which the cell does not respond to new encounters with messenger molecules.

Broadly speaking, this process can be thought of as implementing relay broadcasting: "if a message is received, relay it to one of your nearest neighbours". Hence, though the spatial range of a single messenger molecule may be limited (Shvartsman et al., 2001), the presence of the feedback loop may allow messages to be transmitted over long distances, very much like the spreading of epidemics. For instance, in a detailed deterministic continuous biophysical model, Pribyl et al. (2003a) showed that this kind of relay autocrine signalling may give rise to messenger travelling waves that spread over the entire cell population.

However, like many processes in cell biology, this mechanism comes with inherent noise or stochasticity at several levels. First, because of the diffusive nature of the messenger movements, the target cell of an emitted messenger is random, i.e. cannot be precisely specified. Secondly, because it relies on intrinsically stochastic biochemical reactions, the triggering of an intracellular signalling pathway upon cell-messenger molecule interaction is probabilistic. Finally, at every moment, a messenger molecule can be removed from the system, either by extracellular proteolysis, or by scavenging in the extracellular space.

In recent years, several papers have presented biophysical models for auto/paracrine signalling. However, to the best of our knowledge, these studies consisted either in deterministic models of auto/paracrine relay transmission (Pribyl et al., 2003a,b) or in stochastic models for messenger-cell interactions, but without feedback relay loops (Batsilas et al., 2003; Shvartsman et al., 2001).

The goal of the present work is thus to study the influence of noise in auto/paracrine-like relay broadcasting systems. In particular, it investigates the collective behaviour exhibited by the mutual coupling between cells and their environment, and how messages can be transmitted in stochastic conditions. In this framework, we focused on the basic mechanisms underlying these generic processes, and voluntarily restricted our model to the elementary ingredients identified above. The results demonstrate that classical deterministic modelling fails to predict the behaviour observed in stochastic simulations because of large fluctuations due to the intrinsic noise. Moreover, we show that the behaviour of this stochastic system is due to the occurrence of a nonequilibrium phase transition and present evidence that this nonequilibrium phase transition is in the universality class of directed percolation.

2 The Model

As already mentioned above, we are interested here in the most fundamental processes implied in auto/paracrine-like relay signalling. Hence, we study a "minimal" model that incorporates only the few basic ingredients enumerated above (messenger diffusion, feedback-positive loop for messenger relay, refractory phase). Furthermore, the model is intrinsically stochastic regarding cell stimulation by messenger molecules, as well as messenger molecule survival in the extracellular space. For the same reasons, we study a one-dimensional version of the model, i.e. messenger molecules and cells evolve in a one-dimensional lattice of linear size L with periodic boundaries (i.e. a circle made of L equidistant lattice sites).

2.1 *Messenger Molecules*

Messenger molecules undergo isotropic diffusion on the lattice via non-interacting nearest-neighbour random walks without excluded volume conditions (i.e. several messengers can occupy a same site at a given time). Every messenger can also be removed from the lattice with rate λ per time step.

2.2 *Cells*

Besides messenger molecules, the lattice also contains n immobile point cells that are regularly scattered over the lattice. Each cell occupies a single lattice site, so that, as $n < L$, the distance between two nearest-neighbour cells is several lattice sites. Each cell $i = 1 \ldots n$ is associated with a dynamical variable $\phi_i(t)$ that takes integer values between 0 and $K - 1$ and represents the state of cell i at simulation time step t. Upon stimulation by a messenger molecule, a cell enters a refractory phase which lasts $K - 1$ time steps and during which the cell is not responsive to messenger molecules (see Fig. 1). The state $\phi_i = 0$ is the quiescent state. The transition from $\phi_i = 0$ to $\phi_i = 1$ represents the stimulation step and occurs with probability ω (per time step) if a messenger molecule is present on cell i lattice site. Transition from the state $\phi_i = j$ to $\phi_i = j + 1$ then occurs with probability 1 (at each time step) for $j = 1 \ldots K - 1$. After K steps, the cell is back in the quiescent state i.e. goes from $\phi_i = K - 1$ to $\phi_i = 0$ with probability 1. Finally, the transition from $\phi_i = K_p$ to $\phi_i = K_p + 1$ ($K_p \in 1 \ldots K - 1$) is accompanied by the release of a new messenger on cell i lattice site.

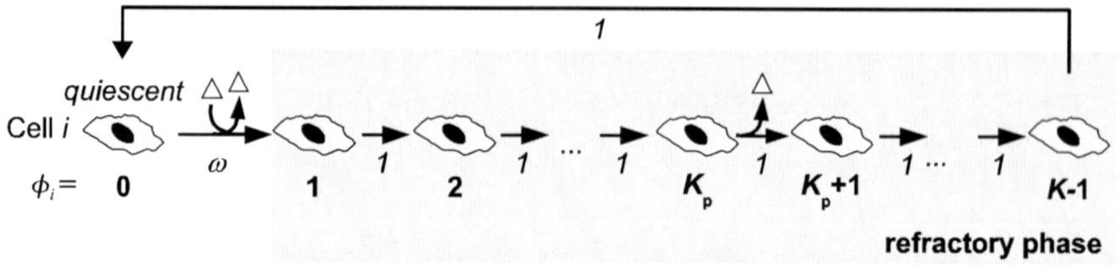

Fig. 1 Overview of the cell activation cycle. In the sequence, the cell state ϕ_i is indicated by a bold number below the cell. Transition probabilities (at each time step) between two cell states are indicated by italicized numbers below corresponding arrows. Intervening messenger molecules are symbolized by white triangles. During the refractory phase of the cycle (*light grey* background), the cell is not sensible to messenger molecules

2.3 Initialization

The initial state of the system corresponds to "full lattice" initial conditions: one messenger molecule is initially positioned *on each* lattice site and the state of each cell $\phi_i(0)$, $\forall i = 1 \ldots n$, is randomly chosen from a uniform distribution between 0 and $K - 1$.

2.4 Simulations

Numerical simulations basically implement the above listed rules (Fig. 1, Sections 2.1. and 2.2). For clarity, we present below the translation of these rules in the point of view of the simulation algorithm.

After initialization, the system is updated as follows at each simulation step $t > 0$:

- The state ϕ_i of each cell i is first updated. If a messenger molecule is present on cell i lattice site and if i is quiescent ($\phi_i(t-1) = 0$), its state is updated to $\phi_i(t) = 1$ with probability ω. Note that, in the presence of m messengers on the same cell site, the effective stimulation probability is $P(\phi_i = 0 \rightarrow \phi_i = 1) = 1 - (1-\omega)^m$. Cells in the refractory phase ($\phi_i(t-1) > 0$) are updated according to $\phi_i(t) = \phi_i(t-1) + 1$ or $\phi_i(t) = 0$ if $\phi_i(t-1) = K - 1$. Finally, new messengers are created at each cell site for which the updated state variable $\phi_i(t) = K_p + 1$.
- The messenger molecules are then updated according to two substeps. Each messenger is first independently and simultaneously moved to one of its two nearest-neighbour sites (chosen at random). Each molecule is then removed from the lattice with uniform probability λ.

2.5 Parameter Values

Unless otherwise stated, standard parameter values used in the present paper were: $L = 500 \times 10^3$ sites, $n = 50 \times 10^3$ cells, $K = 20$ and $K_p = 10$. For each parameter set, the results presented are averages over 10 different realizations of the initial conditions and of the simulation run. The system behaviour was mainly studied through variations of the stimulation probability ω and the messenger removal probability λ.

3 General Behaviour

3.1 Mean-Field Analysis

To predict the behaviour of the model, a first approach neglects the inherently stochastic nature of the model and assumes that the fluctuations of the number of messenger molecules on the lattice are not significant. This so-called "mean-field" approach fundamentally approximates the model as a *deterministic* process, using differential equations that are actually similar to the mass-action laws used in classical (bio)chemical kinetics (see Berry (2003) for further details). Under these assumptions, the system is predicted to asymptotically reach a steady state (i.e. a stable fixed point). The messenger density $\rho\,(= m/L$ where m is the total number of messenger molecules on the lattice) in the steady state is given by:

$$\rho^{ss} = \begin{cases} 0 & if\ \omega < \omega_c \\ \dfrac{1}{K-1}\left(\dfrac{1}{\omega_c} - \dfrac{1}{\omega}\right) & if\ \omega \geq \omega_c \end{cases} \tag{1}$$

with

$$\omega_c = \frac{\lambda L}{n} \tag{2}$$

Hence, mean-field predictions state that, if $\omega < \omega_c$, messenger molecules are eventually cleared from the lattice (while every cell eventually becomes quiescent). In a way, this corresponds to the death of the system and will be referred to as the "absorbing phase". Conversely, if $\omega \geq \omega_c$, messenger molecules persist indefinitely on the lattice. This phase will be referred to as the "active phase".

When ω is close to ω_c, a Taylor series expansion at first order of $(1/\omega_c - 1/\omega)$ is $(\omega - \omega_c)/\omega_c{}^2$. Thus Eq. (1) can be rewritten as:

$$\rho^{ss} \propto \omega - \omega_c, \quad \text{for} \quad \omega \to \omega_c \text{ in the active phase} \tag{3}$$

Figure 2 presents simulation results illustrating the occurrence of the absorbing and active phases. In this figure, the x-axis is the 1-D lattice. A black dot is drawn at each lattice site containing a messenger molecule. The evolution of the system in time is plotted along the y-axis. When ω is small (Fig. 2A), messenger molecules persist for circa 100 time steps, before undergoing massive extinction. Eventually, after $\approx$600 time steps, all the messengers have been removed from the lattice, and

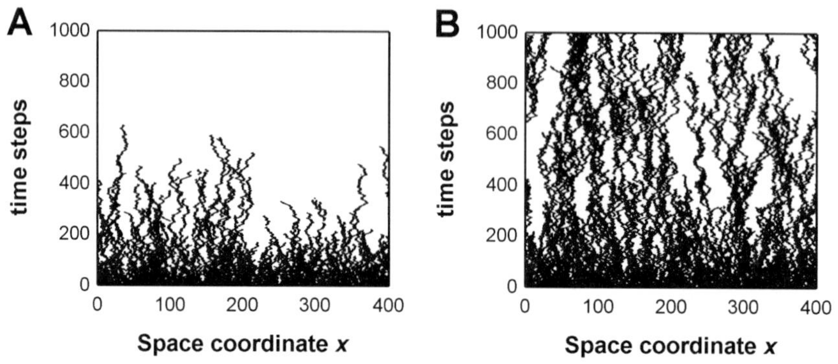

Fig. 2 Simulations of the model illustrating the occurrence of an absorbing (**A**) and an active (**B**) phase. Parameters were: $L = 400$ sites, $n = 40$ cells, $K = 20$, $Kp = 10$, $\lambda = 0.0150$ and (**A**) $\omega = 0.100$ or (**B**) $\omega = 0.6235$

the system freezes into the absorbing (dead) phase. Conversely, for large ω values (Fig. 2B), the messenger molecules survive for the entire time span of the simulation (active phase). Albeit the system is constantly changing, visual inspection of the figure indicates that messenger density tends to a roughly constant value.

It must however be emphasized that the mean-field analysis above is valid only if the spatial fluctuations of messenger molecules are negligible. Yet simulations of the model such as those presented in Fig. 2 show that at every time step, the distribution of messenger molecules is rather inhomogeneous in space, with "white" zones (devoid of messengers) of various sizes and "black" regions where messenger density is high. This is a fundamental expression of the inherently stochastic nature of the model. It indicates that spatial fluctuations in the system are high, and could invalidate some of the predictions made by the *deterministic* mean-field analysis. To quantify and characterize the behaviour of the system with its natural stochasticity and fluctuations, intensive simulations of the model are needed. The following of the paper is devoted to these analyses.

4 Simulation Results

4.1 Dynamics of the Messenger Density

Figure 3 presents the time evolution for the density of messenger molecules ρ in the active (Fig. 3A) or absorbing (Fig. 3B) phases. In the active phase, ρ decreases according to three main regimes. After an initial slow decay regime (up to ≈ 20 time steps), ρ decreases abruptly for $\approx 10^3$ time steps. Finally, at long times, ρ dynamics reaches a regime where it fluctuates around a constant average value. Fundamentally, this stationary fluctuating regime is the stochastic equivalent of the steady state predicted by the deterministic mean-field analysis (Eq. 1). The average value of the fluctuations at long times in the simulations will thus as well be noted ρ^{ss}.

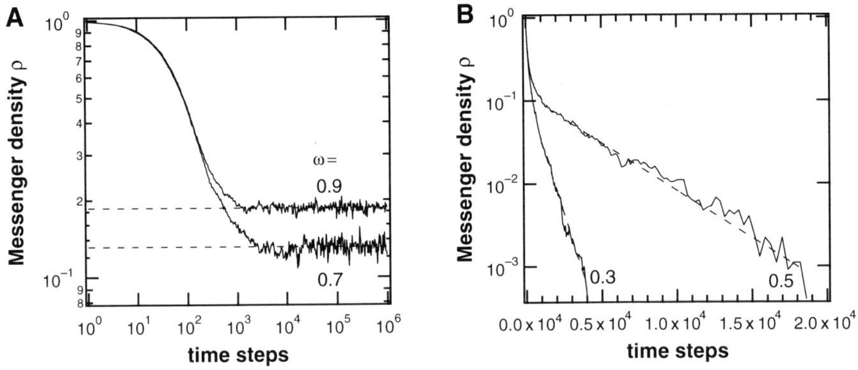

Fig. 3 Dynamics of the messenger density $\rho(t)$. Simulation results for "Full lattice" initial conditions in the active (**A**) or absorbing (**B**) phases. In (**A**) the dashed lines indicate the corresponding average value in the fluctuating stationary state ρ^{ss}. In (**B**) dashed lines represent fits with exponentially decreasing functions. The values of ω are indicated beside the curves; $\lambda = 0.0150$; $L = 500 \times 10^3$ sites, $n = 50 \times 10^3$ cells, $K = 20$ and $K_p = 10$. Results are from a single simulation run

Note however that, while both simulations and mean-field analysis indicate a stationary state in the active phase at long times, the predicted values are different. For instance, with the parameters of Fig. 3, Eq. (1) predicts $\rho^{ss} = 0.276$ (for $\omega = 0.7$) and 0.292 (for $\omega = 0.9$) while simulations yield 0.131 and 0.186, respectively. This is a first evidence for failure of the mean-field analysis due to the high fluctuations in the model.

In the absorbing phase (Fig. 3B), the behaviour at long times is quite contrasted. Instead of reaching a stationary regime, messenger density vanishes very fast. Indeed, this curve is plotted in log-linear coordinates so that the observed linear behaviour indicates an exponentially fast decay.

Hence, in the active phase at long times, ρ settles onto a fluctuating stationary regime, while in the absorbing phase, messengers vanish according to an exponential decay. The difference between these two behaviours can be used to identify the boundary between the two phases. Figure 4A shows the dynamics of ρ for values of ω that are close to the transition between active and absorbing phases. For the smallest values of ω (the bottom curves in the figure), the curves bends down. Note that the curves are plotted here in log-log coordinates so that this downward curvature reflects the exponential decay characteristic of the absorbing phase. Conversely, the curves obtained for the largest values of ω (the upper curves in Fig. 4A) inflect upwards and tend towards the stationary state, which is a characteristic of the active phase.

At the boundary between the two phases, i.e. for the critical value ω_c, the decay of ρ at long times is thus expected to be linear in log-log coordinates (neither curving upwards nor downwards), which corresponds to the power-law:

$$\rho(t) \propto t^{-\alpha}, \quad \text{for} \quad \omega = \omega_c, \text{ and } \quad t \to \infty, \tag{4}$$

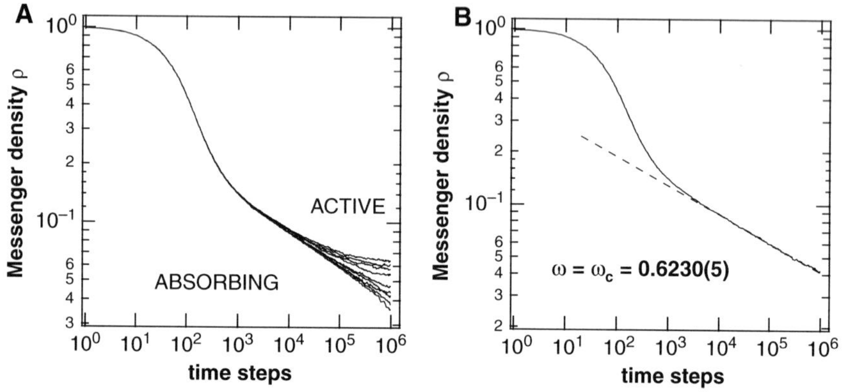

Fig. 4 Dynamics of the messenger density $\rho(t)$ close to the boundary between active and absorbing phases. (**A**) Each curve corresponds to a different value of ω, i.e., from bottom to top, $\omega = 0.6220, 0.6225, 0.6230, 0.6235, 0.6240, 0.6260, 0.6270, 0.6280$ and 0.6290. (**B**) Time-variation of ρ when ω is set to its estimated critical value $\omega = \omega_c = 0.6230(5)$. (The number between parentheses denotes the estimated incertitude on the last digit). The dashed curve indicates a power-law decay with exponent -0.165. Parameters were $\lambda = 0.0150$ and other parameters from the standard set (c.f. 2.5)

Hence, the value of ω that yields a power-law behaviour can be used as an estimate for ω_c. Using this principle, Fig. 4B shows that an estimate of $\omega_c = 0.6230$ is obtained. Furthermore the estimated value for the exponent of the corresponding power-law in this case is $\alpha \approx 0.165$. It must be noted that the prediction of the

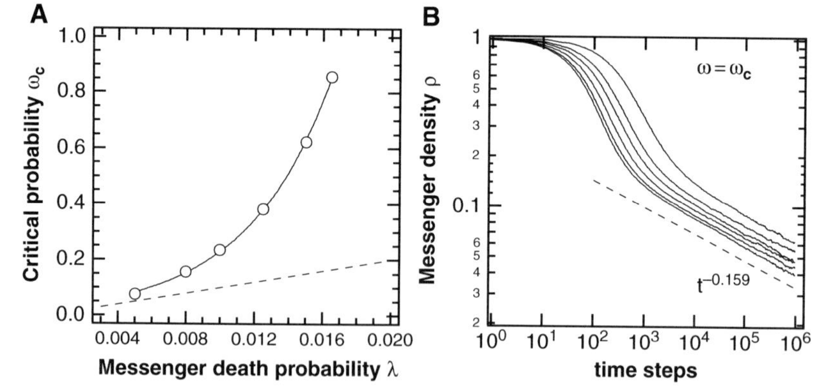

Fig. 5 Critical behaviour for various values of the messenger death probability λ. (**A**) Estimates of the critical stimulation probability ω_c as a function of λ (open circles). The dashed line indicates prediction from the mean-field analysis (Eq. 2). The full line is an exponential fit. (**B**) Dynamics of the messenger density $\rho(t)$ for various values of λ at the corresponding critical value ω_c. λ values are (from *top* to *bottom*): 0.0050; 0.0060; 0.0100; 0.0125; 0.0150 and 0.0165. The dashed line indicates a power-law decay with exponent 0.159. All other parameters according to the standard values (Section 5)

mean-field analysis fails here. In the case of Fig. 4, (Eq. 2) predicts $\omega_c = 0.150$, which is less than one fourth of the value obtained in the simulations.

The critical behaviour for various values of the messenger death (removal) probability λ is shown in Fig. 5. Simulations of the model show that the critical value ω_c increases exponentially fast with λ (Fig. 5A, open circles). Here again, the prediction of the mean-field analysis (Fig. 5A, dashed line) clearly underestimates the actual value of the critical threshold. The dynamics of ρ for various values of λ at criticality (i.e. setting ω to its critical value ω_c at the corresponding value of λ) are plotted Fig. 5B. Albeit the final power-law regimes start at longer times when λ decreases, all the curves obtained eventually reach a power-law regime with similar value of the exponent α (i.e. parallel straight lines in log-log coordinates). Averaging over these simulations, an improved estimate for the exponent α (Eq. 4) is obtained: $\alpha = 0.159(7)$ (the number between parentheses indicates the estimated uncertainty on the last digit).

4.2 A Nonequilibrium Phase Transition

The dependence of the messenger density in the (fluctuating) stationary state ρ^{ss} on the stimulation probability ω is exemplified in Fig. 6 for $\lambda = 0.0150$. The behaviour exhibited by ρ^{ss} in the simulations is clearly typical of a (continuous or second-order) phase transition between the active phase and the absorbing phase. It is for instance similar to the ferromagnetic phase transition observed in materials such as iron, where global magnetization increases continuously from zero as the temperature is lowered below the critical (Curie) temperature[1].

However, a fundamental difference between the phase transition observed here and classical phase transitions lies in its nonequilibrium nature. Indeed, most of the classical phase transitions (including the ferromagnetic transition) are equilibrium phenomena, i.e. they obey the so-called "detailed balance" condition (Hinrichsen, 2006): the transition rate (or more precisely fluxes) from any state i to any state j is equal to that of the reverse transition rate, from j to i. This condition implies the presence of real equilibrium states and is a fundamental property of these systems. In particular, it allows applying usual thermodynamic tools and concepts.

In the system studied here, the absorbing state induces a violation of the detailed balance condition: this state can be reached with some probability from the active state, *but cannot be escaped*. In turn, this prevents the application of several of the classical thermodynamic concepts. The general properties of these kinds of phase transitions, called nonequilibrium phase transitions, are thus still a mater of debate

[1] Note that the apparent discontinuity close to the critical value in Fig. 6 is a numerical artefact due to the so-called "critical slowing down". Indeed, as the critical threshold is approached, the fluctuating stationary regime starts at increasingly long simulation times. As this regime has to be reached to estimate the corresponding value of ρ^{ss}, the number of time steps in the simulations must be substantially increased as the critical threshold is approached. Hence the apparent discontinuity of Fig. 6 only reflects computational limitations hindering measurements with ω values closer to the threshold.

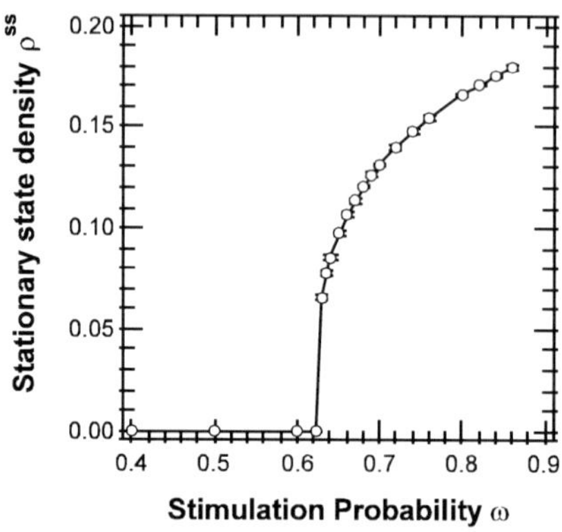

Fig. 6 Variation of the messenger molecule density in the (fluctuating) stationary state, ρ^{ss}, as a function of the stimulation probability ω. Open circles show the results from the simulations of the models (the full line is a guide for the eyes). Parameters were $\lambda = 0.0150$, and standard values for the others. Error bars show ± 1 s.d.

in the physics community, but several points are well established (for reviews see Hinrichsen, 2006, Odor, 2004 or Hinrichsen, 2000a).

Like their equilibrium counterparts, nonequilibrium phase transitions exhibit generic power-law behaviours close to the critical threshold. Equation (4) is a first example of such a power-law, but we will encounter more of them in the following. Several quantities found in the power-laws depend on details of the simulation and the system considered: they are called "non-universal". For instance, the value of the critical threshold ω_c is known to be non-universal, and thus depends on the probability λ for instance, as seen in Fig. 5A (just like the value of the percolation threshold depends on the type of percolation model considered, i.e. bond or link percolation). However, close to the critical threshold, several quantities (called universal quantities) are largely insensitive to microscopic details of the system (as in equilibrium phase transitions). For instance, the value of the exponent α (the critical exponent of power-law Eq. (4) does not vary when λ changes (Fig. 5B).

One great success of the theory of equilibrium phase transitions is the explanation that transitions arising in different systems can share the same set of critical exponents. This phenomenon is called "universality". For example, the critical exponents at the liquid-gas critical point are independent of the chemical composition of the liquid. Furthermore, several models for transport in porous media (Sahimi, 1994) share the same critical exponent values as models of sol-gel transitions (Adam & Lairez, 1996): they are in the same universality class (that of percolation, in this case). This means that the dominant processes implied in these systems are explainable by the same fundamental mechanisms.

Nonequilibrium phase transitions may as well be classified into several universality classes. Among them, the universality class of Directed Percolation (DP) has proven the most robust. Apparently very different systems, ranging from population dynamics (Lipowski & Liposwka, 2000), epidemic spreading (Dammer &

Hinrichsen, 2003), forest fires (Albano, 1994), biological evolution (Ferreira & Fontanari, 2002), Ca^{2+} signalling (Timofeeva & Coombes, 2004; Bär et al., 2000) to morphology dynamics of growing surfaces (Hinrichsen, 2000a), belong to the DP universality class. This indicates that, beyond their diversity, all these systems rely fundamentally on the same ground processes. However, several other universality classes for nonequilibrium phase transitions have been uncovered, such as the parity conserving universality class (Zhong & ben-Avraham, 1995), or the conserved threshold transfer process universality class (Lübeck & Heger, 2003).

Hence, determining the universality class of a nonequilibrium phase transition is an important step towards the understanding of its fundamental mechanisms. Practically, this consists in estimating the value of the critical exponents of the system. This task is carried out in the following sections.

4.3 Universality Class of the Phase Transition

Besides Eq. (4), an important power-law for nonequilibrium phase transitions relates the average value of the density of messengers in the fluctuating stationary state, ρ^{ss}, to the distance to the critical threshold, $\omega - \omega_c$ (Hinrichsen, 2006):

$$\rho^{ss} \propto (\omega - \omega_c)^{\beta}\,, \quad \text{for} \quad \omega \to \omega_c \text{ in the active phase} \tag{5}$$

The corresponding critical exponent β is a universal quantity. Actually Eq. (5) is identical to Eq. (3) with $\beta = 1$, which means that the mean-field analysis predicts $\beta = 1$. Figure 7 presents simulation results showing how ρ^{ss} varies as a function of the distance to the threshold for various values of λ. These curves show roughly parallel straight lines at small distances, indicating a power-law behaviour with a critical exponent that does not depend on λ. Averaging over the 6 conditions of the figure, we obtain an estimate of $\beta = 0.29(1)$. Here again, note that the value predicted by the mean-field analysis is more than threefold the measured one.

Another important critical exponent is the exponent related to the temporal correlations (see Section 5), $\nu_{||}$. Theoretical arguments (Hinrichsen, 2006) indicate that plots of $\rho(t) \times t^{\alpha}$ as a function of $t \times (\omega - \omega_c)^{\nu_{||}}$ should yield a collapse of the curves shown Fig. 4A on two curves: every curves corresponding to active phase conditions should collapse to a single one, while those corresponding to absorbing phase conditions should collapse to another single one. Hence, finding the value of $\nu_{||}$ that yields the best collapse of the curves gives an estimate for this critical exponent.

Figure 8 shows the results obtained for $\lambda = 0.0150$ (i.e. Fig. 4A). Using the estimated value of $\alpha = 0.159$ (see above), the best collapse of the curves (at long times) was found for $\nu_{||} \approx 1.7$. Similar treatments of the data for other values of λ yielded comparable quantities (not shown). On average, the estimate for the critical exponent is $\nu_{||} = 1.66(9)$.

A third important critical exponent is the so-called dynamic exponent z. Roughly speaking, this exponent relates the behaviour of the system to the size of the spatial

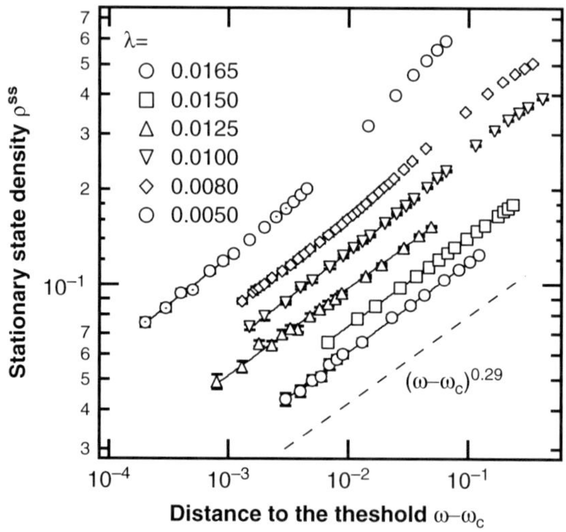

Fig. 7 Scaling of the messenger density in the fluctuating stationary states with the distance to the threshold for different values of the messenger death probability λ. The dashed line indicates a power law with exponent 0.29. All other parameters are set to their standard values. Error bars show ± 1 s.d.

domain (i.e. the number of sites L in the lattice in our case) (Hinrichsen, 2006). It is important to realize that the various power-laws mentioned above are strictly valid for infinite spatial domains only, which is not accessible with computer simulations.

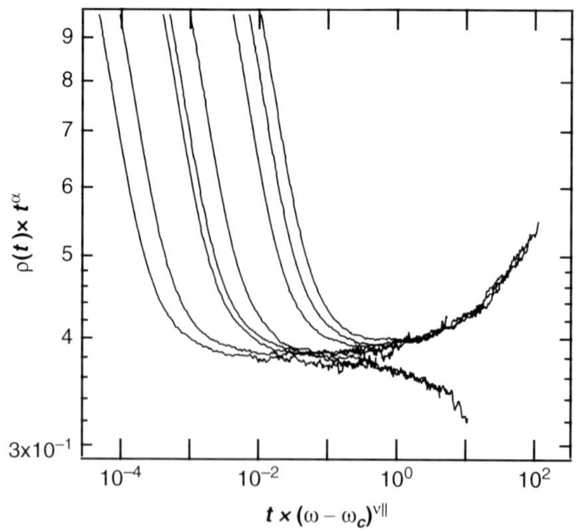

Fig. 8 Scaling test for the value of the critical exponent $\nu_{||}$. The data of Fig. 4A are plotted here as $\rho(t)\times t^{\alpha}$ as a function of $t\times\omega-(\omega_c)^{\nu_{||}}$ with $\alpha = 0.159$. The best collapse of the curves is obtained for $\nu_{||} = 1.70$. Parameters: $\lambda = 0.0150$ and all other parameters according to their standard values

The lattice size used in the simulations presented up to this point ($L = 5 \times 10^5$) was large enough so that the effects of this finite size could be neglected. In fact, for simulation durations of 10^6 time steps, we have found that finite size effects are negligible as soon as $L \geq 5 \times 10^3$ sites.

Smaller L values however evidence drastic effects. Figure 9A shows the dynamics of ρ for small lattices ($L \leq 10^3$) at criticality ($\omega = \omega_c$) for $\lambda = 0.0150$ and constant cell density (i.e. the number of cell is $n = 0.10 \times L$). In this case, the system has always a certain probability to jump into the absorbing phase (the corresponding curves bend downwards) even though the parameters correspond to the critical threshold (of course, this probability decreases when lattice size increases, and becomes very low for large but finite systems). Theoretical arguments (Hinrichsen, 2006) predict that in this case, plots of $\rho(t) \times t^{\alpha}$ as a function of t/L^z should yield a collapse of the curves shown Fig. 9A on a single curve. As with $\nu_{||}$ above, this gives a method for estimating the dynamic exponent z. Fig. 9B shows the best collapse, obtained with $z = 1.55$. The results of Fig. 9 can be replicated using other values for λ, and yield comparable estimates for z (not shown). The average value obtained is $z = 1.54(5)$.

Table 1 summarizes the values obtained for the four critical exponents estimated in this study. It also indicates an estimate for the critical exponent related to the spatial fluctuations, $\nu_{\perp}$. This exponent was not measured directly but deduced from the definition $z = \nu_{||}/\nu_{\perp}$ (Hinrichsen, 2006). The table also indicates the corresponding values for the directed percolation (DP) universality class in one space dimension. Clearly, for each exponent, the obtained estimates are in good agreement with those of DP. This demonstrates that the nonequilibrium phase transition observed in the present model belongs to the directed percolation universality class.

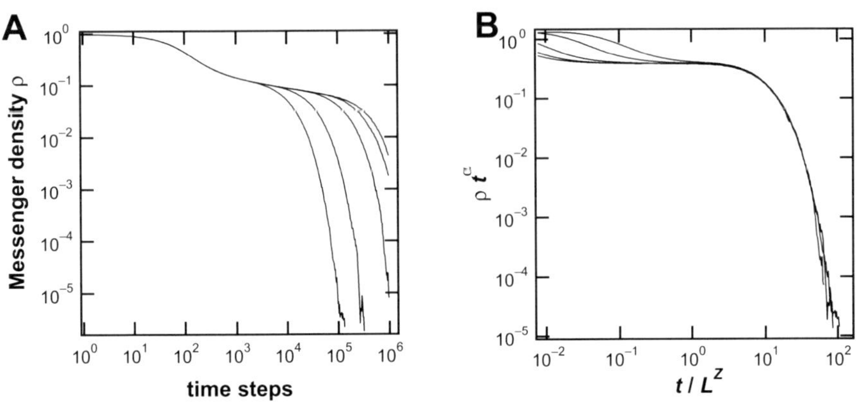

Fig. 9 Finite size scaling test at criticality ($\omega = \omega_c$) and $\lambda = 0.0150$. (**A**) dynamics of the messenger density $\rho(t)$ at criticality for different sizes of the lattice L =100; 200; 500; 800; and 1,000 (from *bottom* to *top*). Corresponding values of the number of cells $n = L/10$ (constant cell density). (**B**) Scaling test for the value of the critical exponent z. The data of (**A**) are replotted here as $\rho(t) \times t^{\alpha}$ as a function of t/L^z with $\alpha = 0.159$. The best collapse of the curves is obtained for $z = 1.55$. Parameters: $\lambda = 0.0150$ and all other parameters according to their standard values

Table 1 Comparison of the estimates obtained for the critical exponents studied in the present model and those of Directed Percolation (DP) in one space dimension (values are from Hinrichsen (2006)). Note that the value of z for the present model is not a measure, but is deduced from $z = \nu_{||}/\nu_{\perp}$. Numbers in parentheses indicate uncertainty in the last digit

Model	α	β	$\nu_{\|\|}$	z	$\nu_{\perp}$
Present	0.159(7)	0.29(1)	1.66(9)	1.54(5)	1.08
DP	0.159464(6)	0.276486(8)	1.733847(6)	1.580745(10)	1.096854(4)

5 Message Broadcasting

The characterization of the phase transition allows a better understanding and prediction of several aspects of the model, such as the characteristics of message broadcasting. Let us imagine that the lattice is in the absorbing (dead) phase, i.e. devoid of any messenger molecule, with all cells quiescent. At time t_0, a single messenger molecule is produced by a cell at a random position on the lattice (see corresponding simulations in Fig. 10). Will this embryo of a message persist over time? Will it finally be transmitted to every cell in the lattice? In physicists' terms, this raises the question of the correlation length in the system. It turns out that, unlike their equilibrium counterparts, nonequilibrium phase transitions possess two independent correlation lengths: a spatial correlation length $\xi_{\perp}$, and a temporal correlation length $\xi_{||}$ (Hinrichsen, 2000a, 2006). These lengths are directly related to the spreading of the message seed.

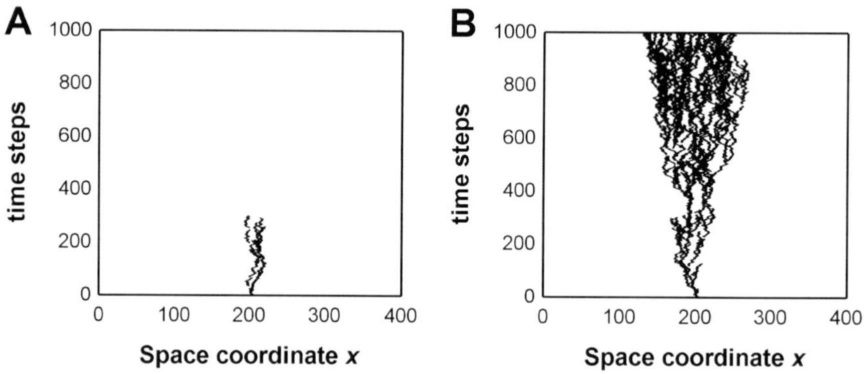

Fig. 10 Simulations of the model illustrating the spreading of a unique initial messenger "seed" in the absorbing phase (**A**) $\omega = 0.50$ or in the active phase (**B**) $\omega = 0.95$. Parameters were: $L = 400$ sites, $n = 40$ cells, $K = 20$, $Kp = 10$, $\lambda = 0.0150$. Each cells were initiated in the quiescent state ($\phi_i(0) = 0$; $\forall i = 1 \ldots n$)

In the absorbing phase, the message survives for a certain amount of time and tends to slowly expand in space (Fig. 10A). The theory of nonequilibrium phase transition tells us that the average survival time is given by (or at least proportional to) $\xi_{||} \propto |\omega - \omega_c|^{-\nu_{||}}$ and the maximal expansion of the message in space by $\xi_{\perp} \propto |\omega - \omega_c|^{-\nu_{\perp}}$ (Hinrichsen, 2000a). Similarly, in the active phase, message spreads as a cone (Fig. 10B) whose opening angle is given by $\xi_{\perp}/\xi_{||}$. Using these theoretical concepts, it becomes easy to understand that, as the critical threshold is approached from below in the simulations, the message spreading from the initial seed survives for increasing times and reaches a growing fraction of the cells. At the critical threshold, the effect of a single seed is even eventually transmitted (relayed) all along the lattice, whatever the physical distance to the initial seed site (both $\xi_{||}$ and $\xi_{\perp} \rightarrow \infty$). As ω increases further (above the threshold) and enters the active phase, the opening angle of the spread cone grows as $|\omega - \omega_c|^{\nu_{||}-\nu_{\perp}}$. Thus, as $\nu_{||} - \nu_{\perp} = 0.637 > 0$ in DP, the time needed for the message to invade the whole lattice decreases with increasing ω. Hence identifying the phase transition allows predicting both qualitatively and quantitatively the behaviours shown by the simulations.

6 Discussion

The idea that the stochasticity inherent to cellular processes has a strong influence on elementary behaviours such as toggle switch properties (Tian & Burrage, 2006; Vilar et al., 2003), clock oscillations (Suel et al., 2006) or signal transduction (Bhalla, 2004), has recently emerged as a new paradigm in cell biology (for a review, see Rao et al., 2002). Furthermore, the role played by noise and stochasticity at the cell or organism level is beginning to be realized (for recent reviews see Samoilov et al., 2006 or Raser & O'Shea, 2005).

The results of the present work add auto/paracrine-like relay signalling to the list of basic cell processes that may be highly influenced by noise-induced fluctuations. Our main objective was to study the influence of noise in a model for stochastic communications where cells respond to a diffusive environment that they themselves have produced through a positive feedback loop. Using numerical simulations, we showed that the deterministic (mean field) approximation of this process fails short of predicting the behaviour of the system. Instead, the latter could be understood as a nonequilibrium phase transition that was evidenced to be in the universality class of directed percolation.

That the model belongs to the directed percolation universality class is actually not totally unexpected. Indeed, DP has been conjectured (Grassberger, 1982) to be the universality class of models exhibiting a phase transition from a fluctuating active phase to a unique absorbing phase; whose dynamic rules involve only short-range processes; and that have no unconventional attributes such as additional symmetries (the actual conditions are in fact a bit more restrictive, see Hinrichsen, 2000a). The present model verifies these conditions thus supporting this conjecture.

Modelling studies of auto/paracrine relay transmission have rarely been addressed so far. In two studies, Pribyl and coworkers (Pribyl et al., 2003a,b) proposed a mechanistic model that presents similarities with the model studied here. Their model takes messenger diffusion, reversible binding to cell surface receptors and a positive feedback from ligand binding to new ligand release. The main differences with our model are that the feedback mechanism in their model is slightly more detailed (from a biochemical perspective) and most notably, their model is *purely deterministic*. Hence, from a stochastic point of view, this model could be considered as a mean-field limit. These authors show that their model supports travelling wave solutions that spread throughout the cell layer. In these waves, the front connects a steady state with nonzero messenger density to a steady state devoid of messengers. These behaviours are at first sight reminiscent of the spreading waves encountered in directed percolation, for instance for "seed" initial conditions (see Section 5), and of the active and adsorbing stationary states. Whether this model is related to directed percolation is however not trivial and would necessitate further works. For instance, the spreading waves in DP are not travelling waves (i.e. the shape of the wave front changes during its propagation). Starting from the discrete version of their model (Pribyl et al., 2003b), biologically realistic stochastic processes could be introduced at several points. Possible phase transitions and universality class of the resulting model could then be assessed through the study of the time evolution of the probability $P(t)$ that a spreading waves survives up to time t, that scales in the subcritical regime as $P(t) \sim t^{-\delta}$ with $\delta = 0.1595$ or 0.451 for DP in $d = 1$ or 2 dimensions, for instance.

Biological interpretation of the present work must be handled with great care. Indeed, the aim of the present work was to study the basic mechanisms at play in such kinds of diffusive communications, so that we voluntarily restricted the model to a set of a few elementary ingredients. Of course, this limits the biological realism of the model, and raises the question of the persistence of the observed behaviours in more detailed and biologically plausible models. While this question can only be answered by further modelling and simulation works, we give here some remarks related to this point.

First of all, the universality class of a phase transition is, by definition, not changed by details of the model. And directed percolation, in particular, has proven a very robust universality class to this respect. Hence, it seems unlikely that minor modifications in the model (such as changing the lattice geometry or using off-lattice conditions, using stochastic refractory periods for the cells...) would modify drastically the observed behaviour. For instance, preliminary investigations showed that changing the values of the cycle parameters (K and K_p) modifies the value of the critical threshold but not the transition itself. Actually, shorter cycle lengths K (not shown) decrease the duration of the initial phase i.e. the non power-law regime of ρ dynamics (the first 10^3 time steps in Fig. 4B).

An important parameter is the Euclidean dimension of the lattice. We choose here to study a one-dimensional model because 1-D models are much cheaper in computation time, thus allowing intensive simulations. However, biological realism would rather impose two-(or three-)dimensional spaces. Actually a two-dimensional version of the model presented here has previously been studied in Berry, (2003). However, because simulations in this previous paper were not refined enough, the universality class could not be determined unambiguously. Hence, future work will focus on intensive simulations of the model in two and three space dimensions.

Furthermore, another significant assumption in the present work is the regular spacing of the cells on the lattice. However, preliminary simulations (not shown) indicate that random cell positioning does not modify the occurrence of the nonequilibrium phase transition, but has the deleterious effect that the density of messenger molecules decays as a power-law in the absorbing phase (and not exponentially, as observed for regular spacing)[2]. This invalidates the estimation of the critical threshold based on the criterion of regime change between subcritical and critical conditions, that was used in Fig. 4, for instance. While more computationally expensive, other methods based on finite-size effects for instance (Ortega et al., 1998; Binder & Heermann, 1997) can however be applied in this case and will be used in future studies of the model with random cell positioning.

A surprising point concerning directed percolation is that in spite of its amazing robustness in models/simulations, this critical behaviour has still not been evidenced in experiments (see Hinrichsen, 2000b). Hence, the possibility that autocrine relays may provide experimental evidence of DP should be considered seriously. Possible experimental setups could for instance consist in cell and tissue culture assays, where a confluent (or subconfluent) layer of autocrine cells is covered by a liquid medium, in which the soluble messenger is secreted and diffuses. Alternatively, such a situation is naturally found in *Drosophilia* eggs (see Pribyl et al., 2003b). The evolution of messenger concentration in the liquid or of related reporters could be monitored in situ, for instance using fluorescence. Such in situ recordings of spatio-temporal evolutions are already accessible to cell biologists (see e.g. Nikolic et al., 2006). Now, messenger death probability (λ in the present model) could be modulated through the addition of various levels of proteinases (degrading the messenger) in the liquid medium. Likewise, the stimulation probability ω could be modulated through the addition of various levels of antagonists of the messenger receptor, or intracellular inhibitors of the signalling pathway supporting the positive feedback. Finally, the A431 carcinoma cell line could be a good candidate for these studies (Dent et al., 1999).

References

Adam M., & Lairez, D., 1996, Sol-gel transition. In *Physical Properties of Polymeric Gels* (J.P. Cohen Addad, ed.), John Wiley &. Sons, UK, pp. 87–142.

Albano, E.V., 1994, Critical behaviour of a forest fire model with immune trees, *J. Phys. A* **277**: L881–L886.

Bär, M., Falcke, M., Levine, H., & Tsimring, L.S., 2000, Discrete stochastic modeling of calcium channel dynamics, *Phys. Rev. Lett.* **84**:5664–5667.

Batsilas, L., Berezhkovskii, A.M., & Shvartsman, S.Y., 2003, Stochastic model of autocrine and paracrine signals in cell culture assays, *Biophys. J.* **85**:3659–3665.

Berry, H., 2003, Nonequilibrium phase transition in a self-activated biological network, *Phys. Rev. E* **67**:031907.

[2] Note that this phenomenon has already been encountered in the literature, see Szabo et al. (2002) for instance.

Bhalla, U.S., 2004, Signaling in small subcellular volumes. I. Stochastic and diffusion effects on individual pathways, *Biophys. J.* **87**:733–744.

Binder, K., & Heermann, D.W., 1997, *Monte Carlo Simulation in Statistical Physics*, Springer, Berlin, Germany.

Dammer, S.M., & Hinrichsen, H., 2003, Epidemic spreading with immunization and mutations, *Phys. Rev. E* **68**:016114.

De Kievit, T.R., & Iglewski, B.H., 2000, Bacterial quorum sensing in pathogenic relationships, *Infect. Imun.* **68**:4839–4849.

Dent, P., Reardon, D.B., Park, J.S., Bowers, G., Logsdon, C., Valerie, K., Schmidt-Ullrich, R., 1999, Radiation-induced release of transforming growth factor alpha activates the epidermal growth factor receptor and mitogen-activated protein kinase pathway in carcinoma cells, leading to increased proliferation and protection from radiation-induced cell death, *Mol. Biol. Cell* **10**:2493–2506.

Ferreira, C.P., & Fontanari, J.F., 2002, Nonequilibrium phase transition in a model for the origin of life, *Phys. Rev. E* **65**:021902.

Freeman, M., 2000, Feedback control of intercellular signalling in development, *Nature* **408**:313–319.

Gloster, J., Freshwater, A., Sellers, R.F., & Alexandersen, S., 2005, Re-assessing the likelihood of airborne spread of foot-and-mouth disease at the start of the 1967–1968 UK foot-and-mouth disease epidemic, *Epidemiol. Infect.* **133**:767–783.

Grassberger, P., 1982, On phase transitions in Schlögl's second model, *Z. Phys. B* **47**:365–374.

Hammond, G.W., Raddatz, R.L., & Gelskey, D.E., 1989, Impact of atmospheric dispersion and transport of viral aerosols on the epidemiology of influenza, *Rev. Infect. Dis.* **11**:494–497.

Hinrichsen, H., 2006, Non-equilibrium phase transitions, *Physica A* **369**:1–28.

Hinrichsen, H., 2000a, Nonequilibrium critical phenomena and phase transitions into absorbing states, *Adv. Phys.* **49**:815–958.

Hinrichsen, H., 2000b, On possible experimental realizations of directed percolation, *Braz. J. Phys.* **30**:69–82.

James, S., Nilsson, P., James, G., Kjelleberg, S., & Fagerström, T., 2000, Luminescence control in the marine bacterium *Vibrio fischeri*: An analysis of the dynamics of lux regulation, *J. Mol. Biol.* **296**:1127–1137.

Lipowski, A., & Liposwka, D., 2000, Nonequilibrium phase transition in a lattice prey-predator system, *Physica A* **276**:456–464.

Lübeck, S., & Heger, P.C., 2003, Universal finite-size scaling behavior and universal dynamical scaling behavior of absorbing phase transitions with a conserved field, *Phys. Rev. E* **68**:056102.

Nikolic, D., Boettiger, A., Bar-Sagi, D., Carbeck, J., & Shvartsman, S., 2006, Role of boundary conditions in an experimental model of epithelial wound healing, *Am. J. Physiol. Cell Physiol.* **291**:C68–C75.

Odor, G., 2004, Universality classes in nonequilibrium lattice systems, *Rev. Mod. Phys.* **76**:663–724.

Ortega, N.R.S., Pinheiro, F.S., Tomé, T., & Drugowich de Felicio, J.R., 1998, Critical behavior of a probabilistic cellular automaton describing a biological system, *Physica A* **255**:189–200.

Pribyl, M., Muratov, C.B., & Shvartsman, S.Y., 2003a, Long-range signal transmission in autocrine relays, *Biophys. J.* **84**:883–896.

Pribyl, M., Muratov, C.B., & Shvartsman, S.Y., 2003b, Discrete models of autocrine cell communication in epithelial layers, *Biophys. J.* **84**:3624–3635.

Rao, C.V., Wolf, D.M., & Arkin, A.P., 2002, Control, exploitation and tolerance of intracellular noise, *Nature* **420**:231–237.

Raser, J.M., & O'Shea, E.K., 2005, Noise in gene expression: origins, consequences, and control, *Science* **309**:2010–2013.

Sahimi, M., 1994, *Applications of Percolation Theory*, Taylor & Francis, UK.

Samoilov, M.S., Price, G., & Arkin, A.P., 2006, From fluctuations to phenotypes: the physiology of noise, *Sci. STKE* **366**:re17.

Shvartsman, S.Y., Wiley, H.S., Deen, W.M., & Lauffenburger, D.A., 2001, Spatial range of autocrine signalling: modelling and computational analysis, *Biophys. J.* **81**:1854–1867.

Suel, G.M., Garcia-Ojalvo J., Liberman, L.M., Elowitz, M.B., Tian, T., & Burrage, K., 2006, An excitable gene regulatory circuit induces transient cellular differentiation, *Nature* **440**:545–550.

Szabo, G., Gergely, H., & Oborny, B., 2002, Generalized contact process on random environments, *Phys. Rev. E* **65**:066111.

Tian, T., & Burrage, K., 2006, Stochastic models for regulatory networks of the genetic toggle switch, *Proc. Natl. Acad. Sci. USA* **103**:8372–8377.

Timofeeva, Y., & Coombes, S., 2004, Directed percolation in a two-dimensional stochastic fire-diffuse-fire model, *Phys. Rev. E* **70**:062901.

Vilar, J.M., Guet, C.C., & Leibler, S., 2003, Modeling network dynamics: the lac operon, a case study, *J. Cell Biol.* **161**:471–476.

Wiley, H.S., Shvartsman, S.Y., & Lauffenburger, D., 2003, Computational modeling of the EGF-receptor systems: a paradigm for systems biology, *Trends Cell Biol.* **13**:43–50.

Zhong, D., & ben-Avraham, D., 1995, University class of two-offspring branching-annihilating random walks, *Phys. Lett. A* **209**:333–337.

Interfacial Water Compartments on Tendon/Collagen and in Cells

I.L. Cameron and G.D. Fullerton

Abstract Bulk water exists in three phase states: crystalline solid, liquid and vapor. Phase transitions between these states are brought about by changes in the physical environment such as temperature and pressure. Each phase of water is characterized by differences in translational and rotational motion of the water molecule. For example, water molecules in ice move one million times slower than in the liquid state. Does this adequately describe behavior of water at the multitude of surfaces in the cell? A molecular model of four water of hydration subcompartments on tendon/collagen, on globular proteins and in a cell is presented. Each subcompartment is defined in terms of compartmental protein hydration capacities, compartmental water motional restrictions and related descriptors. The range of water molecule motional activity in biological systems are described by correlation times that range from that of ice at 10^{-6} s to bulk water at 10^{-12} s. The existence and size of water subcompartments with intermediate motional and other properties has been measured by proton NMR titration, differential scanning calorimetry, centrifugal dehydration force flow rate, isotherm rehydration and osmotic behavior. In addition sub-compartment capacities can be predicted by molecular modeling. The molecular model of water subcompartments first developed for tendon/collagen appears generally applicable for globular proteins and for human erythrocytes. However, changes in protein conformation and aggregation cause changes in the size of the outer-most sub-compartments. Such changes bring about changes in the physical and physiological properties of cellular water. This growing awareness of the role of water subcompartments in explaining cellular activities is predicted to have a revolutionary effect on cell biology.

Keywords Water hydration · cells

I.L. Cameron
University of Texas Health Science Center, San Antonio, USA
e-mail: CAMERON@uthscsa.edu

G.H. Pollack, W.-C. Chin (eds.), *Phase Transitions in Cell Biology*,

1 Introduction

Water undergoes well known phase transitions going from ice to liquid to vapor states as temperatures increases. Under standard conditions, these phase transitions occur at well established temperatures. The translational and rotational correlation time of the water molecule decreases from 10^{-6} s in ice to 10^{-12} s in the bulk liquid state. But, what is the state of water on macromolecular surfaces? As water has a dipolar charge it can align in an electric field and hydrogen bond with either positively or negatively charged sites and the translational speed of the water molecule will then be slowed. If the dipolar water molecule is positioned between a positively charged site on the surface of a substrate and a nearby negatively charged site on the substrate separated by about 4–5 Å, the single water molecule forms a bridge between the charge pair. The water molecule is further slowed in its translational and rotational motion. Such a situation results in a "water bridge." If the opposite charge sites are even further apart, three water molecules can form a double water bridge between the charge pair, resulting in motional restriction similar to that of the single water molecule bridge but involving three times as much water. Completion of the hydrogen bonding water molecule network or cluster tied to the immobilized water bridges creates a surface layer of water over hydrophilic surfaces restricted in motion but to a lesser extent than bridge water molecules. Hydrophobic surfaces, because they are non-polar in charge, do not allow hydrogen bonding of dipolar water molecules but rather cause water molecules near the non-polar surface to form a closed pentagonal ring of water molecules to form into a clathrate structure (Urry 1995). Apparently, more stable surfaces allow more than a monolayer of water molecules to form (Ling 2001; Zheng and Pollack 2006). This multilayer of water molecules is also thought to be somewhat slowed from bulk water in its motional activity. Such multilayered water is now known to differ from bulk water in its physical properties, i.e. freezing (Hori 1956),osmotic (Fullerton and Cameron 2006), solute excluding properties (Ling 2001; Zheng and Pollack 2006).

Changes in the free energy of water compartments on the protein surface charges cause changes in the phase transition behavior of various motionally restricted water subcompartments. Furthermore, it seems likely that changes in surface charge, in folding and unfolding of globular proteins and in protein aggregation will cause changes in the size of various motionally restricted water sub-compartments. Such changes can and do result in the amount of motionally perturbed water in outer water sub-compartments and related perturbation of phase transitions. The importance of understanding water phase transitions is just beginning to be appreciated by cell biologists.

Multiple methods indicate that a majority of intracellular water has properties that differ from ordinary bulk (i.e. tap) water (Ling 2001, 2004; Pollack 2001; Cameron et al. 1997, 2007b; Fullerton and Cameron 2006). This non-bulk-like water in cells is divisible into subcompartments with different degrees of motional restrictions (Cameron et al. 1988, 1991, 2007b). As cellular proteins constitute the vast majority of dry mass in most cells they are known to provide a large proportion of solute interfacial surface, it is therefore logical to look at protein surfaces as the major site for water molecule interactions and motional restrictions. To get at the interaction of water molecules at the structural level of cellular proteins it was first decided to focus attention on a motionally stable protein with a well-established molecular structure, i.e. tendon/collagen. The specific structural interactions with water might therefore be defined at the molecular level. Hopefully the information obtained could then be applied to globular proteins *in vitro* and *in vivo*.

How is motional restriction defined and what accounts for the motional restriction of water molecules? The following description is summarized from Fullerton and Amurao 2006, Fullerton and Rahal 2007 and Fullerton et al. 2006a. Basically it is the free energy relationships of water molecules bound between charge sites on the protein backbone that are held apart by the stiff proline and amide bonds in the molecular chain. By slowly adding water to a dry protein, the water first avidly accumulates in "single-water bridges" at the lowest free energy level (Cameron et al. 2007a). On protein the first water bridge is between a positive charged amide site and a nearby (about 5Å) carbonyl site. If oppositely charged sites on the protein are 3Å or less apart then direct bonding of the two counter-charge sites occurs and a water bridge will not reduce the electrostatic energy of a direct hydrogen bonding.

How restricted is a water molecule in a "water bridge"? Given that water in bulk state rotates or translocates every 6.7×10^{-12} s (correlation time) and water in an ice crystal is slowed to about 10^{-6} s, there is a million-fold slowing of water molecule motion in ice compared to the bulk water state. For comparison, a globular protein molecule in dilute solution, because of its size, moves, rotates or translocates about once every 10^{-9} s, or a thousand-fold slower than does a water molecule in the bulk water state. This description is, however, incomplete. The molecule in the water bridge behaves like a solid molecule along the axis of the bridge but rotates freely like a liquid in the remaining two dimensions. This causes orientational variations of water proton NMR relaxation times as discussed in the next section.

2 Water Sub-Compartments on Tendon/Collagen

Now let us return to the question of movement of a bridged water molecule. The best measures indicate a movement of about 10^{-9} s on tendon/collagen or about the same correlation time as the much larger globular protein. As the protons on the water molecule in a water bridge and the protons of a globular protein in dilute aqueous solution have the same motional restriction is it not always easy to distinguish difference in motional characteristics between the proton populations of water and those of globular protein using proton NMR relaxation time measures. However, use of a relatively immobile protein like that of type 1 collagen in a tendon, which is relatively stable, does allow for measurement of "water bridge" motion using proton NMR relaxation time measures at relatively low hydration levels of the tendon/collagen. Measurement of NMR proton relaxation times at decreasing levels of dehydration allows distinction of three to four different water compartments with different correlation times and sizes and degrees of proton motional restriction (Fig. 1, from Fullerton et al. 2006a).

The sizes and motional properties of the water molecule compartments on tendon/collagen are listed in Table 1.

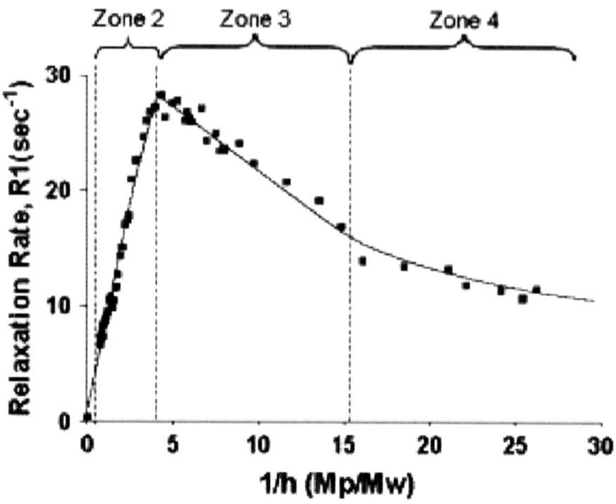

Fig. 1 Proton NMR relaxation rate versus reciprocal of hydration of tendon. Multiple linear segments over the range of hydration from native to dry protein are indicated. This method demonstrates the ability to explain a molecular model of water of hydration fractions. The break points in the data indicate three compartments (Zones 2, 3 and 4) that are summarized in Table 1

The correlation times range from 10^{-9} s for water bridges to 10^{-10} s for main chain cleft water to 10^{-11} s for orientational water to 10^{-12} s for bulk water. As reported in Table 1 the largest fraction of tendon/collagen is orientational water which can be divided into two subfractions. One fraction is ascribed to polar surfaces on the collagen and a second fraction to non-polar surfaces (Fullerton and Rahal 2006, Cameron et al. 2007a). On native tendon these two surface areas are equal in area and account for about 0.8 g water/g dry mass each. Thus, a monolayer of water on native tendon/collagen amounts to 1.6 g water/g dry mass, which is all the water present in native tendon. If the tendon is washed in distilled water or a normal saline solution the tendon swells and more than a monolayer of water of hydration begins to occur. A centrifugal dehydration force (CDF) method indicates that an incomplete

Table 1 Molecular model of tendon/collagen hydration (from Fullerton and Rahal 2006)

Compartment	Zone	g-water/g-dry mass	Correlation time (s^{-1})[a]
Ramachandran water bridge	4	0.065	10^{-9} (1000 times slower than bulk)
Protein main chain hydration	3	0.26	6.55×10^{-10} (100 times slower than bulk)
Primary (polar) hydration	2	0.8	
Secondary (non-polar) hydration	2	0.8	2.6×10^{-11} (4 times slower than bulk)
Native monolayer hydration		1.6	
Multilayer hydration[b]		Variable	
Bulk water (this report)[b]	1	Variable	6.7×10^{-12}
Total sample		Variable	

[a] For reference, ice is 10^{-6} s (1000000 times slower than bulk)
[b] Water washed; Not present on native tendon

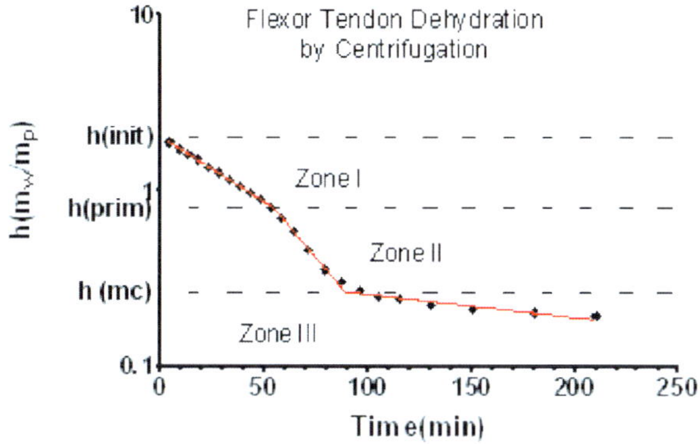

Fig. 2 Results of a new centrifugal dehydration force method for determining multiple water of hydration compartments on water-washed bovine flexor tendon/collagen. The semi-log plot of g water/g dry mass (m_w/m_p) as a function of centrifugation time exposed to a 14000 × g force. The three zones indicate three water flow rate compartments with sizes that closely match predictions by the molecular hydration model in Table 1

second layer of non-bulk water starts to form in the swollen tendon (see Fig. 2 from (Cameron et al. 2007a). A small bulk water fraction is also detected in non-filter paper blotted but water washed tendon (not shown). Thus, multiple water compartments with different correlation times and different thermodynamic parameters are present in the tendon that are different from the properties of water in the bulk phase.

Proton NMR relaxation times done at decreasing levels of hydration and the centrifugal dehydration force procedure clearly indicate the presence of three water compartments with different correlation times. But are these results relevant to globular proteins *in vitro* and *in vivo*? Can tendon/collagen serve as a Rosetta Stone to understand the extent and structural interactions of water molecules in globular proteins and the protein-rich intracellular environment? The direct answer to this question is "yes."

3 Observations on the Extent of Bound Water and Water of Hydration of Globular Proteins *in vitro* and *in vivo*

Protein chemists and cell physiologists commonly accept a water of hydration, or "bound water" value of about 0.3 g water/g dry mass, as summarized in Table 2. All water outside this "bound water" value has been thought to be bulk-like in its physical properties.

This accepted bound water value is far less than the water of hydration values for a growing list of measures on globular proteins in dilute solution. These new measures range from 1.4 to 3.0 g/g or 5 to 10 times more water of hydration than commonly thought to exist on globular proteins (Fullerton et al. 1986, Cameron et al. 1988; Fullerton and Cameron 2006; Fullerton et al. 2006a,b).

Multiple water compartments like those found in tendon/collagen are also found on globular protein such as human hemoglobin (Tables 3 and 4 , from Cameron et al.

Table 2 Extent of "bound water" on proteins and in cells

Specimen	Methods	Capacity (gH2O/gDM)	Reference
Ovalbumin	1–5	0.22 ± 0.07	(Ling 1972)
Serum Albumins	1–8	0.42 ± 0.08	(Ling 1972)
β-Lactoglobin	1,2,5–6	0.41 ± 0.11	(Ling 1972)
Hemoglobin	1,2,4–8,10	0.39 ± 0.02	(Cameron et al. 1988)
Lysozyme	8 different methods	0.267 ± 0.01	(Fullerton et al. 1986)
Frog muscle	Sorption~70–90% rh	0.2	(Ling and Negendank 1970)
Packed Erythrocytes	10	0.22 ± 0.08	(Cameron et al. 1988)

1. Dielectric dispersion; 2. X-ray scattering; 3. Sedimentation velocity; 4. Sedimentation equilibrium; 5. Diffusion coefficient; 6. Intrinsic viscosity; 7. NMR; 8. Frictional coefficient; 9. ^{18}O diffusion; 10. Proton NMR dehydration

1988). Similarly sized water compartments have also been reported on lysozyme (Fullerton et al. 1986, 2006a).

The total water of hydration for human hemoglobin *in vitro* averages 2.82 g/g but in an intact human erythrocyte the value averages only 1.31–1.44 g/g as listed in Table 4 (Cameron et al. 1988). The decrease of water of hydration and of bound water for human hemoglobin *in vivo* has been attributed to aggregation of hemoglobin molecules and loss of water (solvent) accessible surface area as hemoglobin molecules dock to one another. Aggregation of hemoglobin molecules within an erythrocyte has been demonstrated (Cameron et al. 1988, 1991).

The spatial distribution of proteins and other non-cellular molecules in the crowded cytoplasm of an erythrocyte is illustrated by Goodsell in Fig. 3 (from http://mgl.scripps.edu/people/goodsell/).

Such a crowded cytoplasm is a common feature of most cell types (Goodsell 1991). Aggregation of protein molecules may in fact help account for reports of a fraction of bulk water in at least some cells. The existence of some bulk water

Table 3 Comparison of size of multiple water compartments in bovine tendon/collagen and in human hemoglobin A (g water/g dry mass)

Water compartment	Tendon/Collagen			Hemoglobin (# of methods in parentheses)
	Model	NMR	CDF[a]	
Ramachandran water bridge	0.065	0.066	0.062[b]	0.09 (1)
Main chain	0.26	0.26	0.26	0.39 ± 0.02 (6)
Primary hydration	0.8	ND	0.79	0.8 (1)
Secondary (non-polar hydration)	0.8	ND	0.80	0.79 (1)
Native monolayer	1.58	1.6	1.6	1.4 ± 0.05 (4)
Multiple hydration	ND	ND	2.39	2.06 ± 0.05 (2)
Total water of hydration	ND	ND	2.39	2.82 ± 0.17 (7)
Bulk water	ND	ND	0.16	ND

[a] Water washed tendon used for centrifugal dehydration force (CDF) but not for other two methods
[b] Water sorption method
ND = not detected

Table 4 Size of bound and total water of hydration of hemoglobin in vitro and in vivo (g water/g dry mass)

		In vitro	In vivo
I.	*Total water of hydration*	2.82	1.31–1.44
	Bound water	1.31–1.44	0.53–0.55
II.	*Osmotically unresponsive water*	1.7	1.61
	Total water	> 2.82	2.04

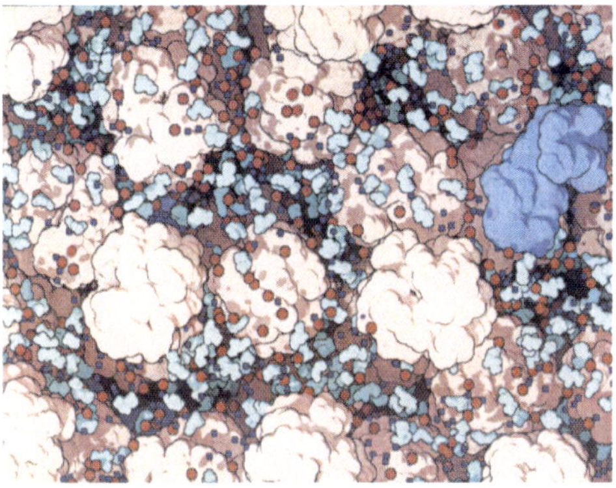

Fig. 3 Crowded cytoplasm of human red blood cell (erythrocyte). This scale drawing by Goodsell shows hemoglobin molecules (*pink*), a host of sugar and amino acids (*green*) and ions (*red*). For clarity, the 530,000 water molecules that would fill all vacant spaces in the image have been omitted. As the dimensions of the hemoglobin molecule were from X-ray crystals it seems likely that hemoglobin molecules would be somewhat larger under the aqueous conditions of a living erythrocyte. This would result in even more crowding than illustrated in the drawing

channels may serve as higher speed fluid pathways for more rapid movement of solutes and perhaps macromolecules from one part of the cell to another location.

4 Conclusion

The physical properties of the different water of hydration subcompartments on proteins and in cells are just beginning to be realized and appreciated. Such properties include for example: non-freezing fractions, osmotically unresponsive water, solute exclusion, flow, viscosity, density, conductivity, diffusion and other properties.

References

Cameron, I.L., Ord, V.A., Fullerton, G.D., 1988, Water of hydration in the intra- and extra-cellular environment of human erythrocytes. *Biochemistry & Cell Biology* 66(11):1186–1199.

Cameron, I.L., Cox, L.A., Liu, X.R., Fullerton, G.D., 1991, Maintenance and mobility of hemoglobin and water within the human erythrocyte after detergent disruption of the plasma membrane, Journal of Cellular Physiology 149:362–372.

Cameron, I.L., Kanal, K.M., Keener, C.R., Fullerton, G.D., 1997, A mechanistic view of the non-ideal osmotic and motional behavior of intracellular water. *Cell Biology International* 21(2):99–113.

Cameron, I.L., Short, N., Fullerton, G.D., 2007b, A centrifugal dehydration force method for characterizing water compartments in fresh and postmortem fish muscle. *Cell Biology International* 31: 516–526.

Cameron, I.L., Short, N., Fullerton, G.D., 2007a, Verification ofsimple hydration/dehydration methods to characterize multiple water compartments on tendon type 1 collagen. *Cell Biology International* 31: 531–536.

Fullerton, G.D., Amurao, M.R., 2006, Evidence that collagen and tendon have monolayer water coverage in the native state. *Cell Biology International* 30(1):56–65.

Fullerton, G.D., Cameron, I.L., 2006, On the osmotically unresponsive water compartment in cells. *Cell Biology International* 30(1):74–77.

Fullerton, G.D., Ord, V.A., Cameron, I.L., 1986, An evaluation of the hydration of lysozyme by an NMR titration method. *Biochimica et Biophysica Acta* 869:230–46.

Fullerton, G.D., Rahal, A., 2007, Collagen structure: the molecular source of tendon magic angle effect. *Journal of Magnetic Resonance Imaging* 25: 345–361.

Fullerton, G.D., Nes, E., Amurao, M., Rahal, A., Krasnosselskaia, L., Cameron, I.L., 2006a, An NMR method to characterize multiple water compartments on mammalian collagen. *Cell Biology International* 30(1):66–73.

Fullerton, G.D., Kanal, K.M., Cameron, I.L., 2006b, Osmotically unresponsive water fraction on proteins: Non-ideal osmotic pressure of bovine serum albumin as a function of pH and salt concentration. *Cell Biology International* 30(1):78–85.

Goodsell, D.S., 1991, Inside a living cell. *Trends in Biochemical Sciences* 16:203–206.

Hori T., 1956, No. 62, US Army, Snow, Ice and Permafrost Res. Establishment. *Low Temperature Science* A15:34.

Ling, G.N., 1972, Hydration of macromolecules. In *Water and Aqueous Solutions: Structure, Thermodynamics and Transport Processes* (R.A. Horne, ed.), Wiley Interscience, New York, USA, pp. 663–700.

Ling, G.N., 2004, What determines the normal water content of living cells? *Physiological Chemistry & Physics and Medical NMR* 36:1–19.

Ling, G.N., 2001, *Life at the Cell and Below-Cell Level: the Hidden History of a Fundamental Revolution in Biology*. (G.N. Ling, ed.), Pacific Press, New York, USA.

Ling, G.N., Negendank, W., 1970, The physical state of water in frog muscle. *Physiological Chemistry & Physics & Medical NMR* 2:15–23.

Pollack, G.H., 2001, *Cells, Gels and the Engines of Life: A New Unified Approach to Cell Function*. (G.H. Pollack, ed.), Ebner & Sons, Seattle, USA.

Urry, D.W., 1995, Elastic biomolecular machines: synthetic chains of amino acids, patterned after those in connective tissue, can transform heat and chemical energy into motion. *Scientific American* 272: 64–69.

Zheng, J., Pollack, G.H., 2006, Solute exclusion and potential distribution near hydrophilic surfaces. In *Water and the Cell* (G.H. Pollack, I.L. Cameron, D.N. Wheatley, eds.) Springer Dordrecht, The Netherlands, pp. 165–174.

The Role of Ion-Exchange on Trypsin Premature Activation in Zymogen Granules

Y.X. Ding, E. Chen, K. Yang and W.-C. Chin

Abstract Sustained elevation in cytosolic Ca^{2+} concentration ($[Ca^{2+}]_C$) is believed to induce premature trypsin activation and initiate acute pancreatitis. Among many triggering factors, ethanol abuse is the most prevalent in provoking acute pancreatitis but with unknown etiological mechanism. Here we show that elevated $[Ca^{2+}]_C$ is directly responsible for inducing premature trypsin activation in the zymogen granules (ZGs) of pancreas acinar cells and is exacerbated by ethanol treatment. *In vitro*, a highly cooperative ion exchange mechanism has been found to regulate elevation in $[Ca^{2+}]_C$ induced premature trypsin activation. The sustained rise in $[Ca^{2+}]_C$ triggers K^+ influx into the ZGs in exchange for the release of matrix bound-$Ca^{2+}$and -H^+. Such K^+/Ca^{2+} ion-exchange mechanism promotes the amplification of mobile-Ca^{2+} concentration in the ZGs ($[Ca^{2+}]_G$) while K^+ replaces bound-H^+ by K^+/H^+ ion-exchange to decrease the pH within ZGs (pH_G). Increasing $[Ca^{2+}]_G$ with concomitant lowering of pH_G generate premature trypsin activation in ZGs. Hence, our data suggest that ethanol abuse provokes a sustained rise in $[Ca^{2+}]_C$ by which K^+ influx induced ion-exchange mechanism initiates premature trypsin activation in acinar ZGs during acute pancreatitis.

Keywords Intracellular Ca^{2+} · ethanol · ion exchange · zymogen granules · trypsin activation · acute pancreatitis

1 Introduction

Pancreatitis, a disease associated with severe abdominal pain, is characterized by inflammation of the pancreas, which potentially leads to pancreatic exocrine and endocrine deficiencies (Mergener and Baillie, 1998). Acute pancreatitis, a sudden severe inflammation of the pancreas, is caused by autodigestion which yields acute abdomen pain. Damages to the pancreas can be elicited by gallstone, alcohol abuse or hypercalcaemia (Mergener and Baillie, 1998). In the United States alone, more

W.-C. Chin
School of Engineering, University of California, UC Merced, Merced, CA, 95344, USA
e-mail: wchin2@ucmerced.edu

G.H. Pollack, W.-C. Chin (eds.), *Phase Transitions in Cell Biology*,

than 50,000 patients are hospitalized annually with an overall mortality of 10–15%. Such statistics escalates to 30% when combined with a multitude of medical complications such as renal or respiratory failure (Gardner et al., 2006). However, after centuries' discovery of the disease and despite much research and many clinical studies, there is still no specific pharmacological therapy for acute pancreatitis.

As a major endocrine/exocrine secretory organ in human body, the pancreas secretes various digestive enzymes into the gastrointestinal tract to facilitate digesting large macromolecules. It is generally believed that the premature activation of these enzymes, including trypsin, plays a pivotal role on the pathogenesis of acute pancreatitis in its initiating phase. Trypsin, a proteolytic digestive enzyme, is synthesized in endoplasmic reticulum and stored in the ZGs within the apical area of the pancreatic acinar cells (Gorelick et al., 1992). After being secreted and targeted to the small intestine, trypsin is activated by brush border hydrolase enterokinase. This initial activation of trypsin can further activate trypsinogen turning into active trypsin and other zymogens, such as chymotrypsinogen, protelastase, and prophospholipase into their active states (Kassell and Kay, 1973; Gorelick and Otani, 1999). Trypsin is also capable of undergoing massive autoactivation before entering the small intestine. In diseased state, abnormal premature activation of trypsin can induce parallel cascades of other premature digestive enzyme activation inside the pancreas which attributes to acute pancreatitis (Steer, 1999; Halangk and Lerch, 2004).

Intracellular Ca^{2+} regulates many cellular activities including the stimulus-secretion coupling in pancreatic acinar cells. Transient Ca^{2+} changes induced by hormones and neurotransmitters control the normal secretion and level of digestive enzymes in the pancreas (Frick et al., 1997). Sustained elevation of $[Ca^{2+}]_C$ has been reported in both experimental and clinical acute pancreatitis cases. As a result, abnormal $[Ca^{2+}]_C$ elevation has been proposed to initiate the intracellular (premature) activation and release of digestive enzymes from ZGs (Gorelick et al., 1992; Parekh, 2000; Kruger et al., 2000). However, the mechanisms to which elevated $[Ca^{2+}]_C$ leads to the premature activation of trypsin remain largely unknown and thus are the focus of this study. Our previous works demonstrated that Ca^{2+} signaling in goblet and mast cells induced granule swelling needed for enzyme release through K^+ ion exchange (Nguyen et al., 1998; Quesada et al., 2001). Here we investigate the effects of $[Ca^{2+}]_C$ elevation following ethanol treatment on premature trypsin activation in ZGs of acinar cells. We also demonstrate that a similar ion exchange system which can be exacerbated by ethanol abuse is responsible for sustained elevation of $[Ca^{2+}]_C$ induced trypsin activation in acute pancreatitis.

2 Materials and Methods

2.1 Materials

All probes including Rhodamine 110, bis-(CBZ-L-isoleucyl-L-prolyl-L-arginine amide) dihydrochloride (BZiPAR) ($\lambda_{excitation} = 498$ nm, $\lambda_{emission} = 521$ nm), Calcium Orange-5N-AM ($K_d = 20$ µM, $\lambda_{excitation} = 549$ nm, $\lambda_{emission} = 576$

nm), Rhod-2 AM (K_d = 570 nM, $\lambda_{excitation}$ = 552 nm, $\lambda_{emission}$ = 581 nm), LysoSensor Blue DND-167 (pK_a = 5.1, $\lambda_{excitation}$ = 373 nm, $\lambda_{emission}$ = 425 nm), LysoSensor DND-189 (pK_a = 5.2, $\lambda_{excitation}$ = 443 nm, $\lambda_{emission}$ = 505 nm), and PBFI (K_d = 10 mM, $\lambda_{excitation}$ = 340 nm, $\lambda_{emission}$ = 505 nm) were obtained from Molecular Probes, Eugene, Oregon. Valinomycin and ionomycin were purchased from Calbiochem, San Diego, California. Type IA collagenase, FCCP (carbonyl cyanide 4-(trifluoromethoxy)phenylhydrazone), MES (2-Morpholinoethanesulfonic acid sodium salt), EGTA (ethylene glycol bis (β-aminoethylether)-N, N, N', N'-tetraacetic acid), NMG (monovalent organic cation N-methyl-D-glucamine) and other chemicals were provided by Sigma-Aldrich corporation, St. Louis, MO.

2.2 *Mouse Pancreatic Acinar Cell Harvest and Dye Loading*

Male CBA/J mice (Jackson Laboratory, Bar Harbor, ME) were euthanized by exposure to CO_2 according to the protocol approved by the Florida State University Animal Care Committee. Isolated pancreatic acinar cells were harvested from mouse pancreas with collagenase (Raraty et al., 2000). Intact isolated pancreatic acinar cells were equilibrated at 37 °C for 30 min in Hanks' solution containing 10 μM Rhodamine 110, bis-(CBZ-L-isoleucyl-L-prolyl-L-arginine amide) dihydrochloride (BZiPAR). BZiPAR is a specific substrate for the serine protease trypsin (Raraty et al., 2000; Kruger et al., 1998). This non-fluorescent substrate turns into fluorescent Rhodamine 110 after cleavage of the two oligopeptide side chains by trypsin (Leytus et al., 1983; Raraty et al., 2000). Due to the technical difficulties of direct measurement of trypsin activity within the ZGs, the fluorescence intensity of Rhodamine 110 was used to represent the trypsin activity. The labeled cells were washed with Hanks' solution to remove any excess dyes.

Similarly, 5 μM Calcium Orange-5N-AM, 1 μM LysoSensor Blue DND-167 or LysoSensor DND-189, and 5 μM PBFI were loaded to monitor $[Ca^{2+}]_G$, pH_G, and K^+ concentration inside the ZGs ($[K^+]_G$) respectively (Nguyen et al., 1998; Quesada et al., 2001). The labeled cells were utilized to prepare isolated ZGs in the following studies.

2.3 *Granule Preparation and Measurement of pH_G, $[K^+]_G$, $[Ca^{2+}]_G$ and Trypsin Activation Inside the ZGs*

Granules were isolated by adhering to established procedure (Nguyen et al., 1998; Ding et al., 2006). Briefly, while still suspended in an intracellular solution (140 mM potassium glutamate, 20 mM Tris, 5 mM $MgSO_4$,10 mM MES, and 2 mM EGTA, pH 7.3) that mimics the intracellular environment, the labeled cells were disrupted by brief sonication with ZGs isolated by centrifugation. The collected granules were then allowed to attach to polylysine-coated chambers filled with intracellular buffer at designed recipes according to different experiments. Free

Ca^{2+} concentration in the intracellular solution ($[Ca^{2+}]_i$) was calculated by using procedures published elsewhere (Nguyen et al., 1998; Kao, 1994) to represent $[Ca^{2+}]_C$. Valinomycin (10 μM, K^+ ionophore), ionomycin (10 μM, Ca^{2+} ionophore), or FCCP (10 μM, H^+ ionophore) were used to equilibrate the K^+, Ca^{2+} or H^+ across ZGs membrane respectively. All studies were performed on the labeled ZGs directly, instead of the whole cells, to eliminate interferences from the cytoplasm or other cellular organelles, and overcome possible motion artifacts generated as a result of monitoring fluorescence inside the ZGs in intact cells (Nguyen et al., 1998; Quesada et al., 2001; Ding et al., 2006). The chambers were mounted and kept at 37 °C on the thermo-regulated stage of an inverted Olympus fluorescence microscope using a 100 × 1.35 NA oil-immersion objective. Fluorescence images were taken and sent through a 12 bit 1392 × 1040 CoolSNAP-Pro CCD camera to a computer equipped with digital deconvolution system (Image-Pro Plus, SharpStack, Media Cybernetics, Silver Spring, MD, USA). The collected fluorescent images were then analyzed to obtain intensity data.

2.4 Effects of Ethanol Treatment on ZGs

To study the effects of ethanol, isolated ZGs were incubated for 2 h at room temperature with intracellular solution containing ethanol (0 ~ 100 mM). $[Ca^{2+}]_i$ varied from 0 to 700 nM to represent the changes in $[Ca^{2+}]_C$ under stimulation (Ding et al., 2006). The changes in pH_G, $[Ca^{2+}]_G$ and trypsin activity within the ZGs were measured using the same method.

2.5 Effects of Ethanol Treatment on Pancreatic Acinar Cells

In order to investigate the role of $[Ca^{2+}]_C$ in ethanol (100 mM) induced trypsin activation, intact acinar cells were loaded with high-affinity Ca^{2+}-probe, Rhod-2, to monitor $[Ca^{2+}]_C$ change (Nguyen et al., 1998).

2.6 Data Analysis

Each experiment was conducted at least three times to capture at least 30 single ZG images. The fluorescence intensity of BZiPAR, indicating trypsin activity, is measured in an arbitrary unit (AU). For Calcium Orange-5N, Rhod-2, LysoSensor DND-167 or LysoSensor DND-189, and PBFI, the fluorescence intensities were calibrated and converted to $[Ca^{2+}]_G$, pH_G, and $[K^+]_G$ respectively using methods described previously (Nguyen et al., 1998; Quesada et al., 2001; Ding et al., 2006). The data are presented as means ± SD where N represents the number of ZGs collected from each experiment. The statistical significance was determined by using

Student's t-test with GraphPad Prism 4.0 (GraphPad Software, Inc. San Diego, CA). Differences were considered statistically significance at P-values < 0.05.

3 Results

3.1 Elevated $[Ca^{2+}]_i$ Induced Changes of $[Ca^{2+}]_G$, pH_G, and Trypsin Activity Inside ZGs

The isolated granules were exposed to intracellular buffer containing different $[Ca^{2+}]_i$ levels. The changes of $[Ca^{2+}]_G$, pH_G, and trypsin activity inside the ZGs were monitored (Fig. 1A, B, and C). The magnitude of resting $[Ca^{2+}]_G$ (9.1 ± 0.9 μM, mean ± SD, n = 65), and pH_G (5.5 ± 0.1, mean ± SD, n = 40) were consistent with their dynamic range as found in previous published reports (Quesada et al., 2001; Ding et al., 2006; Petersen and Sutton, 2006). The trypsin activity change was monitored indirectly through measuring BZiPAR fluorescence intensity variation (fluorescence emission from Rhodamine 110). At resting $[Ca^{2+}]_i$ level, the measured BZiPAR fluorescence was 829.0 ± 14.7 (mean ± SD, n = 40).

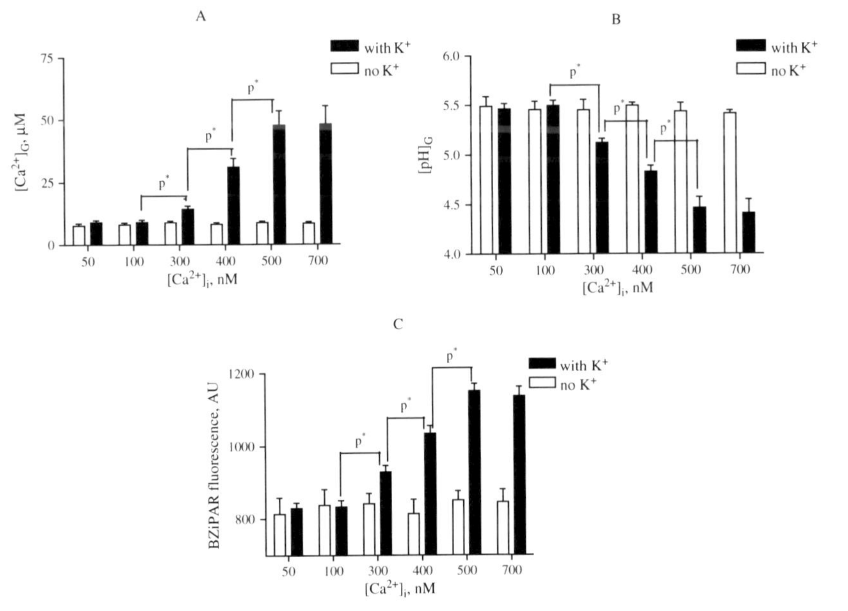

Fig. 1 A–C. Effects of increasing $[Ca^{2+}]_i$ on $[Ca^{2+}]_G$, pH_G and trypsin activity inside the ZGs: **A**. Increasing $[Ca^{2+}]_i$ results in the increase of $[Ca^{2+}]_G$. **B**. Increasing $[Ca^{2+}]_i$ results in the decrease of pH_G. **C**. Increasing $[Ca^{2+}]_i$ results in the increase of BZiPAR fluorescence intensity (trypsin activity) inside the ZGs. $p^* < 0.05$. But, in the absence of K^+, $[Ca^{2+}]_G$, pH_G and trypsin activity inside the ZGs remained the same with the increase of $[Ca^{2+}]_i$, $p^* > 0.12$

According to Petersen and Sutton (2006), the total Ca^{2+} inside the ZGs of pancreatic acinar cells is $\sim$ 15 mM which exceeds either $[Ca^{2+}]_C$ or $[Ca^{2+}]_G$. However, the majority of this measured Ca^{2+} concentration is not in the mobile/free ionic form; instead it is bound to the ZGs matrix (Verdugo, 1990). The mobile Ca^{2+} released from the matrix of isolated granules was measured with membrane permeant dye, Calcium Orange 5N. At physiological resting levels, $[Ca^{2+}]$ of the intracellular solution ($[Ca^{2+}]_i$) remained at around 100 nM. After stimulating with 300 nM of $[Ca^{2+}]_i$, minute changes in $[Ca^{2+}]_G$ was observed. Further increment in $[Ca^{2+}]_i$ to 400 nM yielded a sharp rise in $[Ca^{2+}]_G$. When $[Ca^{2+}]_i$ was increased to 500 nM, the calcium release reached a plateau while additional increment of $[Ca^{2+}]_i$ to 700 nM did not enhance further $[Ca^{2+}]_G$ increase (Fig. 1A). These results suggested that the threshold of $[Ca^{2+}]_i$ for eliciting granule Ca^{2+} dissociation from bound form occurred at around 300–400 nM. Similar data showing H^+ release (pH_G decrease) and BZiPAR fluorescence increase were observed with $[Ca^{2+}]_i$ elevation (Fig. 1B and C). It was also noted that BZiPAR fluorescence increase is highly correlated with Ca^{2+} and H^+ released from the isolated ZGs matrices.

3.2 Ethanol-Induced Changes of pH_G, $[Ca^{2+}]_G$, $[Ca^{2+}]_C$, and Trypsin Activity

$[Ca^{2+}]_i$ was maintained at 500 nM throughout the following experiments. The isolated ZGs were incubated with varying concentrations of ethanol for 120 min. Increasing the concentration of ethanol from 20 to 100 mM exhibited a dose dependent rise in BZiPAR fluorescence emission of 15–20% (Fig. 2A). Our studies also revealed that the sensitization effect of ethanol on $[Ca^{2+}]_i$-induced trypsinogen premature activation followed concentration dependency and was strongly affected by the $[Ca^{2+}]_i$ level (Ding et al., 2006). At resting $[Ca^{2+}]_i$ ($\sim$ 100 nM), treatment of ZGs with ethanol (0–100 mM) failed to alter BZiPAR fluorescence intensity inside the ZGs (data not shown).

Both $[Ca^{2+}]_G$ and pH_G ($[H^+]_G$) are also the triggering factors that initiate trypsin premature activation within the ZGs (Figarella et al., 1988; Gorelick and Otani, 1999). Ethanol incubation was shown to induce changes in $[Ca^{2+}]_G$ and pH_G (Fig. 2B and C). Under normal resting $[Ca^{2+}]_G$ of $\sim$ 10 μM, the application of progressively higher $[Ca^{2+}]_i$ subsequently elicited Ca^{2+} release from the ZGs matrix and boosted $[Ca^{2+}]_G$ to $\sim$ 50 μM (when $[Ca^{2+}]_i = 500$ nM) (Fig 1a). After incubating the ZGs with 100 mM ethanol for 2 h, the $[Ca^{2+}]_G$ further rose to a higher level (Fig. 2B). The 500 nM of $[Ca^{2+}]_i$ also induced the lowering in pH_G from resting $\sim$ 5.7 to below 4.76. Ethanol treatment (100 mM) under same $[Ca^{2+}]_i$ further reduced pH_G from 4.76 to 4.3 (Fig. 2C). In whole cell level, under the stimulation of 100 mM ethanol, we found a long duration of $[Ca^{2+}]_C$ increase immediately after the addition of ethanol (Fig. 2D, $N = 4$), as indicated by rhod-2 fluorescence increase. This result from acinar cells with ethanol exposure was also consistent with a previous report (Gonzalez et al., 2006).

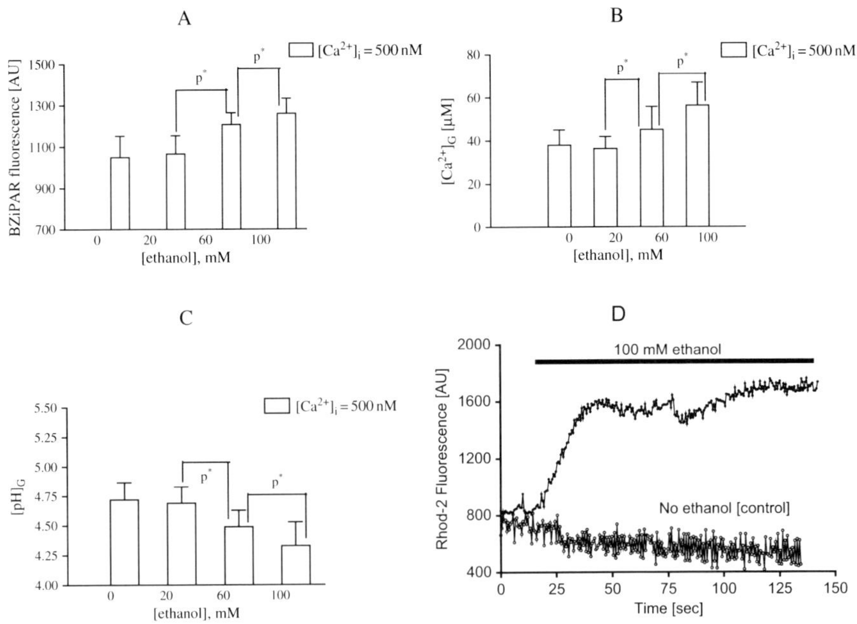

Fig. 2 Effect of ethanol incubation on $[Ca^{2+}]_G$, $[pH]_G$ and trypsin activity inside the ZGs, under 500 nM $[Ca^{2+}]_i$. **A**. BZiPAR fluorescence intensity inside the ZGs (AU), an indicator for trypsin activity inside the ZGs; **B**. $[Ca^{2+}]_G$; **C**. $[pH]_G([H^+]_G)$; **D**. A long duration of $[Ca^{2+}]_C$ increase in intact acinar cells was induced by ethanol and monitored with Rhod-2 as a function of time inside an acinar cell exposed to 100 mM ethanol (N = 4). $p^* < 0.05$. The fluorescence decay of the control is due to photobleach

3.3 K^+/Ca^{2+} and K^+/H^+ Ion-Exchange Inside the ZGs

Cytosolic K^+ influx is the main determinant factor in modulating changes exerted by $[Ca^{2+}]_C$ elevation. In the absence of K^+ influx (K^+-free intracellular solution), increasing $[Ca^{2+}]_i$ (equivalent to $[Ca^{2+}]_C$) has no effect on altering the BZiPAR fluorescence, $[Ca^{2+}]_G$, and pH_G inside the ZGs (Fig. 1). This implies that the sustained elevation of $[Ca^{2+}]_C$ induced trypsin activation relied heavily on the influx of K^+. Figure 3A shows that $[K^+]_G$ elevation can only be initiated with concomitant rise in $[Ca^{2+}]_i$. To explore the mechanism underlying rising $[Ca^{2+}]_C$ induced premature trypsin activation, we investigate whether cytosolic K^+influx could evoke $[Ca^{2+}]_G$ and pH_G changes. Valinomycin was utilized to directly manipulate $[K^+]_G$ inside the ZGs. The effects of varying $[K^+]_G$ on $[Ca^{2+}]_G$ were shown in Fig. 3B. With $[K^+]_G$ below 50 mM, no change in $[Ca^{2+}]_G$was observed; however, there was 4-fold difference in $[Ca^{2+}]_G$ when $[K^+]_G$ approached 100 mM. This indicates that the threshold for $[K^+]_G$ inducing the upsurge in $[Ca^{2+}]_G$ occurred between 50 and 100 mM. As explained in aforementioned section, due to the low $[Ca^{2+}]_C$ (100 nM) at resting physiological condition when compared to $[Ca^{2+}]_G$ level (in μM range), the rise in

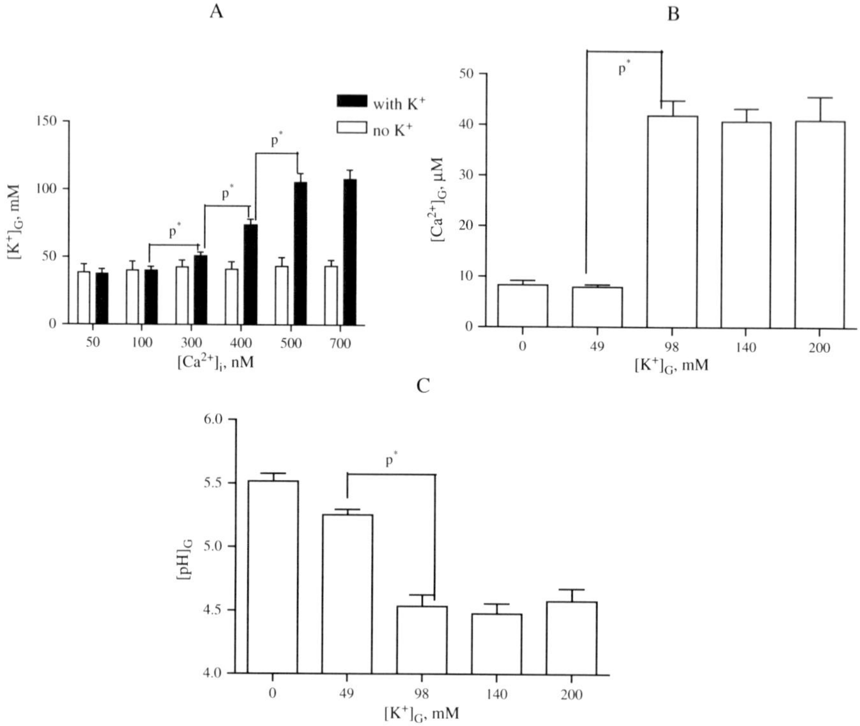

Fig. 3 A–C. K^+ induced ion exchange within the ZGs. The isolated ZGs are equilibrated in the intracellular buffer containing valinomycin. The influx of K^+ to the lumen of ZGs via valinomycin can induce the increase of $[Ca^{2+}]_G$as well as the decrease of $[pH]_G$. $p^* < 0.01$

$[Ca^{2+}]_G$ following elevated $[K^+]_G$ can only be attributed to the released Ca^{2+} that was originally bound to the ZG matrix. Augmenting $[K^+]_G$ also led to a concomitant acidification of ZGs (lower pH) (Fig. 3C).

3.4 The Synergetic Effects of $[Ca^{2+}]_G$ and pH_G on the Trypsin Activation Inside the ZGs

Since both $[Ca^{2+}]_C$ and ethanol incubation could induce changes of $[Ca^{2+}]_G$ and pH_G, it is therefore critical to investigate the possible synergetic effects of $[Ca^{2+}]_G$ and pH_G on the trypsin activation inside the ZGs. FCCP and ionomycin were utilized to equilibrate the H^+ and Ca^{2+} across the ZG membrane respectively. Both the amplified $[Ca^{2+}]_G$ and drop in pH_G lead to the activation of trypsinogen. We also found that the BZiPAR fluorescence intensity was more sensitive to lowering pH_G and culminated only under simultaneous stimulation of increased $[Ca^{2+}]_G$ and decreased pH_G (Fig. 4).

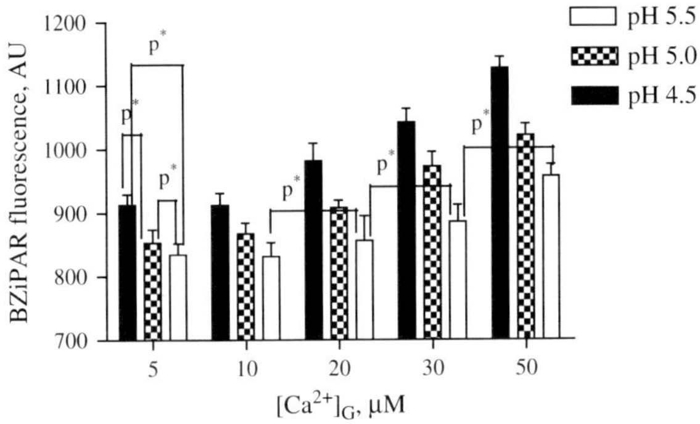

Fig. 4 The synergistic effects of increased $[Ca^{2+}]_G$ and decreased $[pH]_G$ on trypsinogen autoactivation within the ZGs. The isolated ZGs are equilibrated in the intracellular buffer containing 10 μM FCCP and 10 μM ionomycin. Both increased $[Ca^{2+}]_G$ and decreased $[pH]_G$ alone can trigger the increase of trypsin activity inside the ZGs. However, only the combination of increased $[Ca^{2+}]_G$ and decreased $[pH]_G$ can increase trypsin activity inside the ZGs to the maximum. $p^* < 0.05$

4 Discussion

Many pancreatic digestive enzymes are synthesized and stored in the secretory granules as inactive precursor forms that are then activated at proper locations, such as the small intestine. The primary pancreatic digestive enzyme, trypsin, is stored within the ZGs as trypsinogen, an inactive precursor of trypsin. Upon activation, trypsinogen is released from the pancreas and transported to the small intestine where it is proteolytically cleaved by enterokinase to render active trypsin (Kassell and Kay, 1973). Trypsin activation initiates subsequent transformation of other digestive enzymes to their respective active forms in downstream cascade. Under normal physiological conditions, a low level of trypsinogen activation (trypsin activity) exists within the acinar cell (autoactivation) which serves as a protective mechanism that restrains autoactivation of trypsin within the safety limit. The pancreatic secretory trypsin inhibitor (PSTI) blocks trypsin activity by binding to its active site to prevent additional pancreatic enzymes being activated (Kassell and Kay, 1973; Colomb et al., 1979). However, in abnormal diseased conditions these safety mechanisms are inundated by massive trypsin autoactivation and fail to suppress trypsin activity. Such premature trypsin activation is commonly believed to trigger the initiation of acute pancreatitis (Gorelick and Otani, 1999).

Intracellular Ca^{2+} functions as a key regulator for various cellular activities such as secretion of zymogen enzymes from the pancreas acinar cells (Gorelick et al., 1992; Raraty et al., 2000). The canonical pancreatic secretion of digestive enzymes is accomplished by $[Ca^{2+}]_C$changes and is tightly regulated by neurotransmitters or hormones binding to receptors on pancreatic acinar cells (Gorelick et al., 1993; Raraty et al., 2000; Parekh, 2000). Acute pancreatitis begins with a sustained and global elevation in $[Ca^{2+}]_C$. Such rise in $[Ca^{2+}]_C$ has serious consequence that foreshadow the premature activation of trypsin which induces the onset of acute pancreatitis. Our study clearly demonstrated that an upsurge in $[Ca^{2+}]_C$ is directly linked to ZGs stimulation thereby inducing trypsin activation (Fig. 1). In the US, excessive alcohol abuse contributes to more than one third of all cases of acute pancreatitis (Gardner et al., 2006; Mergener and Baillie, 1998). However, the mechanism underlying this major triggering factor on the etiology of acute pancreatitis was unresolved. The data presented here showed that ethanol could sensitize the ZGs to premature zymogen activation induced by $[Ca^{2+}]_C$ elevation (Fig. 2D). Evidently, incubation with ethanol harvested higher BZiPAR fluorescence in comparison to the treatment with $[Ca^{2+}]_i$ alone (Fig. 2A).

To investigate how $[Ca^{2+}]_C$ elevation induces premature trypsin activation in pancreatic ZGs of acinar cell, we hypothesize the involvement of an ion exchange process established previously in goblet and mast cells (Nguyen et al., 1998; Quesada et al., 2001). Firstly, we found the ascending $[Ca^{2+}]_i$ elicits the subsequent rise in $[K^+]_G$ (Fig. 3A). The fact that $[K^+]_G$ elevation induces the increase of $[Ca^{2+}]_G$ as shown by our results (Fig. 3B), authenticate the proposed ion exchange system between K^+ with Ca^{2+}. Secondly, in addition to the K^+/Ca^{2+} ion exchange process, we propose a new K^+/H^+ exchange process occurring within ZGs. In this study, a substantial drop in pH_G (increased $[H^+]_G$) (Fig 3C) with the increment of $[Ca^{2+}]_G$ induced by $[K^+]_G$ rise. The cytosolic influx of K^+ drives the ion exchange process whereby the absence of K^+ inflow prevents the release of the bound-Ca^{2+} and -H^+, thus impeding trypsin activation. This is supported by the consistently low BZiPAR fluorescence throughout the replacement of K^+ in intracellular buffer with NMG (Fig. 1C). Conversely, $[K^+]_G$ rise induces the ion exchange system to release bound-Ca^{2+} and -H^+whose effects are displayed as high free-$[Ca^{2+}]_G$ and declining pH_G (Fig. 3 B and C). The ethanol treatment could also sensitize the ZGs by making the ion exchange processes more effective. After incubation with ethanol, the same level of $[Ca^{2+}]_i$ elevation was observed to induce higher $[Ca^{2+}]_G$ and pH_G changes. The polyanionic mucin polymer matrix found in pancreas acinar ZGs can function as the $K^+/Ca^{2+} - K^+/H^+$ ion exchanger (Fig. 5). Although resting $[Ca^{2+}]_G$ and pH_G adversely affects trypsin activation, the released Ca^{2+} and H^+through ion exchange with K^+influx can favor trypsin premature activation (Figarella et al., 1988; Gorelick and Otani, 1999).

Ethanol treatments were found to enhance acinar cells for zymogen activation induced by CCK or carelum (Katz et al., 1996; Zhao et al., 2002). Our data here corroborated with these observations and further demonstrated that ethanol incubation could sensitize the ZGs to elevated $[Ca^{2+}]_C$ induced trypsin activation (Ding et al., 2006). Previously ethanol was proposed to contribute to the sensitization effects through several mechanisms. First, ethanol may induce sustained global $[Ca^{2+}]_i$ rise (Criddle et al., 2004; Gonzalez et al., 2006; Fig. 2D), initiate serum secretion, or stimulate the release of physiological secretagogue, like CCK (Saluja et al., 1997). Second, ethanol treatment has been shown to affect cellular signaling pathway. Ethanol has been reported to change calcium channel conductivity on plasma membrane, enhance cGMP production, and sensitize G-protein-linked receptors (Dufour and Adamson, 2003). Third, ethanol can change cellular physical structure and function, including membrane fluidity, protein-protein interaction, and lipid structure

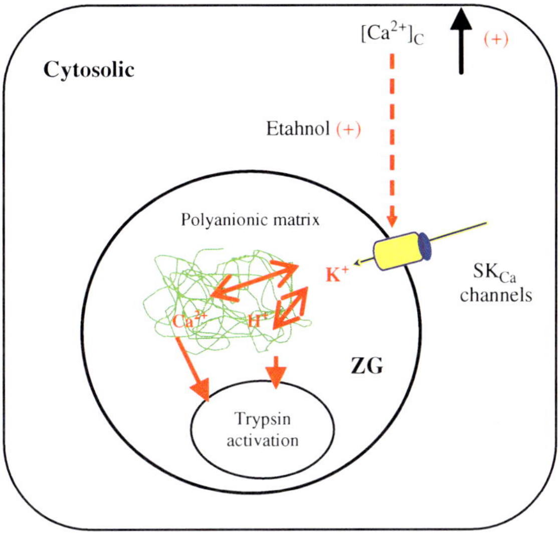

Fig. 5 K^+ induced ion exchange models in ethanol incubation and sustained elevation of $[Ca^{2+}]_C$ induced trypsin premature activation (After Yang et al., 2007) The $[Ca^{2+}]_C$ elevation could elicit the inflow of K^+ into the ZGs. The $[K^+]_G$ rise then launched the ion exchange processes to release bound Ca^{2+} and H^+ from the matrix inside the ZGs. The resulted $[Ca^{2+}]_G$ and pH_G promote the trypsin activation. Ethanol incubation can enhance this process

(Apte and Wilson, 2003). Our current work showed it is more likely that ethanol incubation enhanced trypsin premature activation by inducing the ZGs to be more receptive to $[Ca^{2+}]_C$ induced K^+ ion exchange mechanism. Together with previous observations, these results suggested this proposed ion exchange mechanism could work as a cellular model to investigate the pathophysiology of diseases.

Acknowledgments This work was supported by grants from the Alcoholic Beverage Medical Research Foundation and UC Merced GRC (to W.C.C.).

References

Apte, M.V., and Wilson, J.S. (2003). Alcohol-induced pancreatic injury, Best. Pract. Res. Clin. Gastroenterol. 17, 593–612.

Colomb, E., Figarella, C., and Guy, O. (1979). The two human trypsinogens: Evidence of complex formation with basic pancreatic trypsin inhibitor-proteolytic activity. Biochim. Biophys. Acta. 570, 397–405.

Criddle, D.N., Raraty, M.G., Neoptolemos, J.P., Tepikin, A.V., Petersen, O.H., and Sutton, R. (2004). Ethanol toxicity in pancreatic acinar cells: mediation by nonoxidative fatty acid metabolites. Proc Natl. Acad. Sci. USA 101, 10738–10743.

Ding, Y.X., Yang, K., and Chin, W.C. (2006). Ethanol augments elevated-$[Ca^{2+}]_C$ induced trypsin activation in pancreatic acinar zymogen granules. Biochem. Biophys. Res. Commun. 350(3), 593–597.

Dufour, M.C., and Adamson, M.D. (2003). The epidemiology of alcohol-induced pancreatitis, Pan creas. 27, 286–90.

Figarella, C., Miszcuk-Jamska, B., and Barret, A.J. (1988). Possible lysosomal activation of pancreatic zymogen: activation of both human trypsinogens by cathepsin B ans spontaneous acid activation of human trypsinogen. Biol. Chem. Hoppe-Seyler. 369, 293–298.
Frick, T.W., Fernandez-del-Castillo, C., Bimmler, D., and Warshaw, A.L. (1997). Elevated calcium and activation of trypsinogen in rat pancreatic acini. Gut. 41, 339–343.
Gardner, T.B, Merk, B.S., and Yakshe, P. (2006). Acute pancreatitis. http://www.emedicine.com/med/topic1720.htm.
Gonzalez A., Nunez A.M., Granados M.P., Pariente J.A., and Salido G.M. (2006). Ethanol impairs CCK-8-evoked amylase secretion through Ca2+-mediated ROS generation in mouse pancreatic acinar cells. Alcohol. 38, 51–57.
Gorelick, F.S., Modlin, I.M., Leach, S.D., Carangelo, R. and Katz, M. (1992). Intracellular proteolysis of pancreatic zymogens. Yale J Biol Med. 65, 407–420.
Gorelick, F.S., and Otani, T. (1999). Mechanisms of intracellular zymogen activation. Baillieres. Best. Pract. Res. Clin. Gastroenterol. 13, 227–240.
Halangk, W., and Lerch, M.M. (2004). Early events in acute pancreatitis. Gastroenterol. Clin. North. Am. 33, 717–731.
Kao, J.P.Y. (1994). In: A Practical Guide to the Study of Calcium in Living Cells (ed. Nuccitelle, R.) 155–181 (Academic, San Diego).
Kassell, B., and Kay, J. (1973). Zymogens of proteolytic enzymes. Science 180, 1022–1027.
Katz, M., Carangelo, R., Miller, L.J. and Gorelick, F. (1996). Effect of ethanol on cholecystokinin-stimulated zymogen conversion in pancreatic acinar cells, Am. J. Physiol. 270, G171–G175.
Kruger, B., Lerch, M.M., and Tessenow, W. (1998). Direct detection of premature protease activation in living pancreatic acinar cells. Lab. Invest. 78, 763–764.
Kruger, B., Albrecht, E., and Lerch, M.M. (2000). The role of intracellular calcium signaling in premature protease activation and the onset of pancreatitis. Am. J. Pathol. 157, 43–50.
Leytus, S.P., Melhado, L.L., and Mangel, W.F. (1983). Rhodamine-based compounds as fluorogenic substrates for serine proteinases. Biochem. J. 209, 299–307.
Mergener, K., and Baillie, J. (1998). Acute pancreatitis. B. M. J. 316, 44–48.
Nguyen, T., Chin, W.C., and Verdugo, P. (1998). Role of Ca^{2+}/K^{+} ion exchange in intracellular storage and release of Ca^{2+}. Nature 395, 908–912.
Parekh, A.B. (2000). Calcium signaling and acute pancreatitis: specific response to a promiscuous messenger. Proc. Natl. Acad. Sci. 97, 12933–12934.
Petersen, O.H., and Sutton, R. (2006). Ca^{2+} signaling and pancreatitis: effects of alcohol, bile and coffee. Trends. Pharmacol. Sci. 27, 113–120.
Quesada, I., Chin, W.C., Campos-Bedolla, P., Steed, J., and Verdugo, P. (2001). Mouse mast cell secretory granules can function as intracellular ionic oscillators. Biophys. J. 80, 2133–2139.
Raraty, M., Ward, J., Erdemli, G., Vaillant, C., Neoptolemos, J.P., Sutton, R., Petersen. O.H. (2000). Calcium-dependent enzyme activation and vacuole formation in the apical granular region of pancreatic acinar cells. Proc. Natl. Acad. Sci. 24, 13126–13131.
Saluja, A.K., Lu, L., Yamaguchi, Y., Hofbauer, B., Ruenzi, M., Dawra, R., Bhatia, M., and Steer, M.L. (1997). A cholecystokinin-releasing factor mediates ethanol-induced stimulation of rat pancreatic secretion. J. Clin. Invest. 99, 506–512.
Steer, M.L. (1999). Early events in acute pancreatitis. Baillieres. Best. Pract. Res. Clin. Gastroenterol. 13, 213–225.
Verdugo, P. (1990). Goblet cells and mucus secretion. Annu. Rev. Physiol. 52, 157–176.
Yang, K., Ding, Y.X., Chin, W.C. (2007). K^{+}-induced ion-exchanges trigger trypsin activation in pancreas acinar zymogen granules. Arch. Biochem. Biophys. 459, 256–263.
Zhao, L., Karne, S., Kolodecik, T. and Gorelick, F.S. (2002). Alcohols enhance caerulein-induced zymogen activation in pancreatic acinar cells, Am. J. Physiol. Gastrointest. Liver Physiol. 282 G501–G507.

Whole-Cell Phase Transition in Neurons and its Possible Role in Apoptotic Cell Death

F. Gallyas and J. Pál

Abstract For explanation of the formation of "dark" neurons, a non-enzymatic (physicochemical) mechanism, gel-to-gel phase transition, was recently hypothesized (Gallyas et al., Biol. Cell, 2004; 96:313–324). The present paper pieces together facts that support this mechanism, and draws attention to the possibility that the same physicochemical mechanism also plays a role in the execution phase of apoptosis.

Keywords Degeneration · recovery · mode of death · light- and electron-microscopy

1 Morphological Characteristics of the "Dark" Neurons

It is a century-old, enigmatic phenomenon in human neuropathology that, in the brain of patients who had died of various neurological diseases, the somata, nuclei and main dendrites of a number of neurons are markedly shrunken and hyperchromatic. Such neurons, which are traditionally called "dark", are randomly scattered among normal neurons of the same phenotype, chemistry and function (Fig. 1a). Axons can also become "dark", independently of their parent cell bodies (Fig. 1b).

At present, numerous animal models are available for the production of "dark" neurons and "dark" axons under controlled circumstances by either pathometabolic conditions, such as ischemia or hypoglycemia, or by momentary physical events, such as mechanical or electric injuries. Studies with such models have revealed that all the morphological characteristics of "dark" neurons (Figs. 1–5) have an all-or-nothing nature. This means: if the soma of a neuron is affected, then its dendritic tree is also affected, and *vice versa*; regarding the axons, generally their mm-long segments but not their parent cell bodies or their whole lengths are affected, and *vice versa* (Figs. 2a,b).

It appears impossible that any pathometabolic or physical injury in a given brain area should simultaneously produce the same morphological changes at each point

F. Gallyas
Department of Neurosurgery, University of Pécs, H-7623 Pécs, Rét utca 2, Hungary

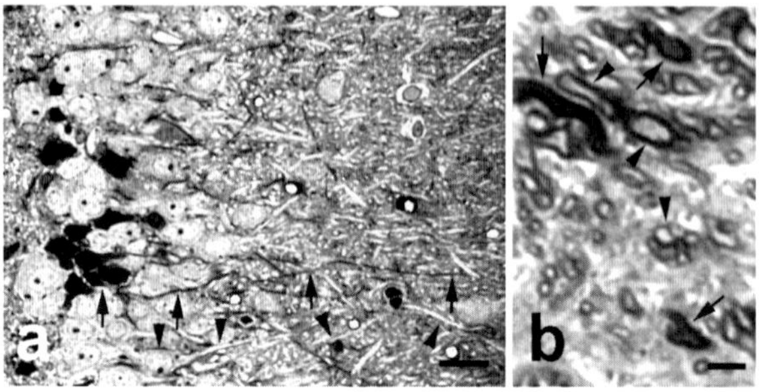

Fig. 1 "Dark" (*arrows*) and normal (*arrowheads*) neurons (**a**) and axons (**b**) in the hippocampal dentate gyrus (**a**) and the caudo-putamen (**b**) of a rat head-injured just before transcardial perfusion-fixation. Toluidine-blue; 1-μm sections; scale bars: **a** = 20 μm, **b** = 5 μm

of a neuronal soma-dendrite domain or a long axon segment, and at the same time should spare all points of many neighboring neuronal soma-dendrite domains or axons, despite the quasi-homogeneity of the initiating noxa. The only reasonable explanation of this phenomenon is that some intracellular perturbation, able to produce the characteristic morphological changes, spreads throughout each affected neuronal soma-dendrite domain or axon segment from a single initiation point.

As a characteristic staining feature, "dark" neuronal soma-dendrite domains and "dark" axon segments bind a massive excess of positively charged dye molecules, such as toluidine blue (Fig. 1). Since this ability can be suppressed by digestion with proteinase-K, but not with nucleases, a striking increase in negative charges in protein molecules must be responsible for this phenomenon (Zsombok et al., 2005).

"Dark" neuronal soma-dendrite domains and "dark" axon segments can be stained black with silver by a reliable (Newman and Jasani, 1998) and selective (Gallyas et al., 2002) argyrophil-III silver method (Gallyas et al., 1990), whereas normal neurons and axons remain unstained (Fig. 2). In the argyrophil-III methods, the formation of each of the multitudinous, discrete, submicroscopic silver grains, whose cumulative light absorption results in the visible staining, is catalyzed at

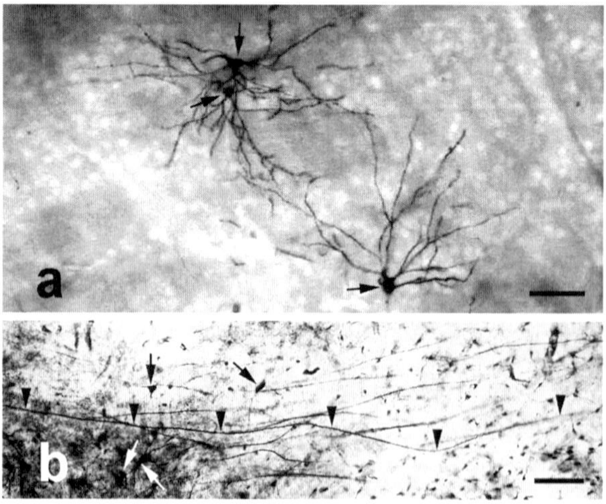

Fig. 2 "Dark" neurons (*arrows*) and "dark" axons (*arrowheads*) in the caudo-putamen (**a**) and the thalamus (**b**) of a rat head-injured just before the transcardial perfusion of an aldehyde fixative. In (**a**), small circles that are lighter than the background indicate the locations of normal neurons. Silver staining; 150-μm sections; scale bars: **a** = 100 μm, **b** = 200 μm

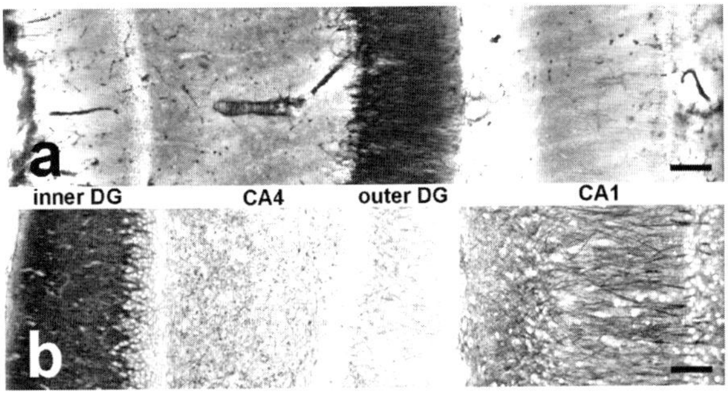

Fig. 3 Electrically produced "dark" soma-dendrite domains of granule neurons in the outer layer of the hippocampal dentate gyrus, which are stained black with silver, are not stained with MAP2 (**a**), whereas the normal granule neurons in its inner layer, which are not stained with silver, are stained with MAP2 (**b**). 50-μm vibratome section; scale bars: 100 μm

sites where a few side groups of protein molecules are present in a favorable spatial arrangement (Gallyas, 1982). Consequently, the stainability with silver by an argyrophil-III method is indicative of the appearance of numerous catalytic sites in protein molecules.

Several antibodies including anti-MAP-2 (Fig. 3) can stain either the soma-dendrite domain or the nucleus in the normal neurons, but not in the "dark" neurons (unpublished observation). This phenomenon is indicative of the disappearance of antigenicity from several kinds of protein molecules in the cytoplasm or the nucleus during the formation of "dark" neurons.

In the electron microscope, a massive increase in electron density and a dramatic ultrastructural compaction, that is a marked decrease in distance between any two neighboring parts of apparently intact ultrastructural elements, can be seen in the cytoplasm and nucleus of every "dark" neuron. The interior of the endoplasmic

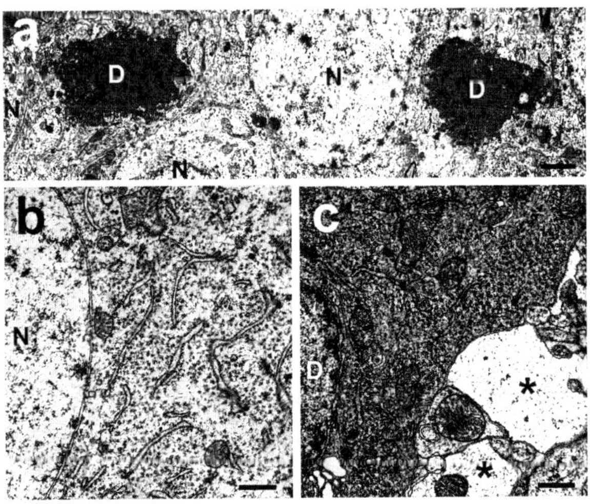

Fig. 4 Hippocampal dentate gyrus of a rat head-injured just before the transcardial perfusion of an aldehyde fixative. D denotes nuclei of "dark" neurons, N nuclei of normal neurons and small asterisks dilated astrocytic processes. Black or white arrowheads point to endoplasmic reticulum cisternae, a white arrow to Golgi cisternae. Scale bars: **a** = 2μm, **b** and **c** = 500 nm

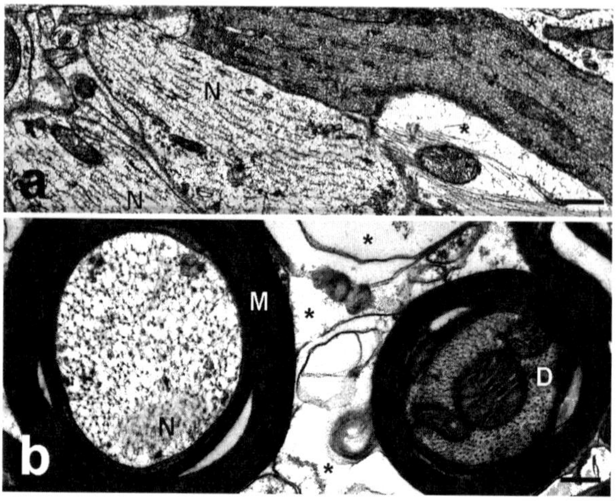

Fig. 5 Normal (N) and "dark" (D) dendrites (**a**) and axons (**b**) in the neocortex (**a**) and caudo-putamen (**b**) of a rat head-injured just before the transcardial perfusion of an aldehyde fixative. M denotes myelin sheaths, asterisks dilated astrocytic processes. Scale bars: **a** and **b** = 500 nm

reticulum cisternae is shrunken, whereas that of the Golgi cisternae is dilated, while the interior of membrane-bound structures such as the mitochondria remains unchanged. Interestingly, the fluid pressed out of the "dark" neuronal somata, dendrites or axon segments is present in nearby astrocytic processes, but not in the extracellular space.

A comparison of normal and "dark" dendrites or normal and "dark" axons reveals more clearly that the ultrastructural elements discernible in the traditional transmission electron microscope remain visibly intact during the formation of "dark" neurons or "dark" axons, and it is the distances between any two of the microtubules in dendrites, or neurofilaments in axons, which are responsible for the increased electron density of the compacted profiles. This suggests that the structural substratum of the ultrastructural compaction that is of the formation of "dark" neurons and "dark" axon segments should be looked for in these spaces. As an equivalent of the light-microscopic all-or-nothing nature, the measure of ultrastructural compaction in every "dark" electron-microscopic profile appears to be the same.

It is an important characteristic of "dark" neuronal elements that all the morphological features demonstrated so far, i.e. the all-or-nothing nature, basophilia, argyrophilia and ultrastructural changes are independent of the event that caused their formation. This is true for not only the physical injuries mentioned (head injury, Gallyas et al., 2004; electric shock, Kellermayer et al., 2006), but also for the pathometabolic conditions tested by us (hypoglycemia, Gallyas et al., 2005; ischemia, unpublished observation), or by other authors (reviewed by Gallyas et al., 2004).

There is a further enigmatic phenomenon to be discussed: By means of the above momentary physical noxae, "dark" neurons and "dark" axon segments displaying all the light- and electron-microscopic features in question can be produced not only *in vivo* but also under various *post-mortem* circumstances: Specifically, in rats chilled down to just above the freezing point by immersion into icy water after a 30-min. transcardial perfusion with an electron-microscopic fixative, or after a 30-min. transcardial perfusion with chilled physiological saline, or after a 5-min. transcardial perfusion with isoosmotic potassium chloride (Gallyas et al., 2004; Kellermayer et al., 2006). Consequently, the formation of "dark" neuronal soma-dendrite domains and "dark" axon segments, that is the process of ultrastructural compaction, can take place even under circumstances extremely unfavorable for enzymatic processes. This argues against any enzyme-mediated mechanism of ultrastructural compaction, as follows: If successive and/or parallel enzymatic processes should cause this process *in vivo*, their harmonized action would be disturbed under the *post-mortem* circumstances tested, which would result in other morphological changes, if any. It is also improbable that any enzyme-mediated process should spread to long intracellular distances from a single point without fading (all-or-nothing phenomenon), or produce numerous negative charges resulting in hyperbasophilia, and catalytic sites for the silver-staining, or cause the disappearance of antigenicity in proteins under the conditions unfavorable for enzymatic processes.

2 Mechanism of the Ultrastructural Compaction in Neurons and Axons

In harmony with the above facts, any mechanism of the ultrastructural compaction taking place in the "dark" neuronal soma-dendrite domains and "dark" axon segments should fulfill the next criteria: (i) It should be a process that can spread to long distances if initiated at a single point (domino principle). (ii) It should be a process that can be initiated by both chemical and physical noxae. (iii) It should be a process that takes place at the expense of stored non-covalent free energy (it does not need any enzyme-mediated energy source). (iv) Its taking place results in the same set of physical and chemical changes, and (v) a dramatic reduction in volume.

It is self-evident that these criteria can not be performed by any fluid, but only by a structure that is evenly and continuously present in whole neuronal soma-dendrite domains or long axon segments. Consequently, this structure cannot be any of the ultrastructural elements "visible" in the conventional electron microscope, since intracellular areas (e.g. nucleus, pericaryon, dendrites, axons, etc.) that comprise different sets of the "visible" ultrastructural elements are affected by the compaction in a similar manner. On the other hand, based on the fact that the synthetic gels capable of phase transition can fulfill each of the above criteria (Tanaka et al., 1985, 1992; Hoffman, 1991; Pollack, 2001, 2002), we hypothesized recently (Gallyas et al., 2004) that this structure consists in an intracellular gel that is continuously present in all spaces among the "visible" ultrastructural elements, and is capable of phase transition.

The hypothesized cytoplasmic gel may be built up as follows: The non-covalent binding sites in most protein molecules that are not components of the "visible" ultrastructural elements are engaged not in maintaining their *in-vitro* conformations but in mutually chaperoning their *in-vivo* conformations, in anchoring the resulting protein matrix to the "visible" ultrastructural elements, in binding potassium ions (Ling and Ochsenfeld, 1973) and in orienting water molecules into multilayers (Ling and Walton, 1975). As regards the energetic requirement of phase transition in this gel, we hypothesized that its non-compacted structure exists in a metastable state of non-covalent free energy, and its non-compacted structure is built up in such a way that the non-covalent free energy, if released by any exogenous activation energy at any intracellular point, inevitably serves as the activation energy at the neighboring points (domino principle).

It is established that the phase transition in a gel consists in a cooperative conformation change in the macromolecules building up its matrix. In our case, as covalent bonds are not involved in the hypothesized gel, such a phase transition can (1) proceed similarly both *in vivo* and *post mortem*, and (2) spread for long distances, beginning from a single point. Furthermore, as a result of the conformational change in protein molecules of the gel matrix, (3) their pre-existing negative groups can be detached from counter-ions (K+?), (4) the spatial arrangement of their pre-existing side groups can be changed favorably for the catalysis of silver staining, and (5) their pre-existing antigenicity can disappear, even under conditions unfavorable for enzymatic processes. Consequently, all the morphological changes characteristic of "dark" neurons argue for the existence of an intraneuronal gel structure capable of whole-cell phase transition. Only a day-long aldehyde fixation can prevent this process by forming covalent bridges between the protein molecules of this gel (see Cammermeyer, 1961).

The widely accepted existence of the plasma membrane and its concomitants (channels, pumps, etc.) appears to be a crucial argument against any gel conception of the whole cytoplasm. However, if the cytoplasmic gel has a spongy-like structure whose confluent cavities are filled with fluid cytoplasm, and the plasma membrane is in contact with a large number of such cavities, then membrane-derived cell functions and gel-derived cell functions could coexist (unpublished presumption).

3 Relationship of the Formation of "Dark" Neurons to the Process of Apoptosis

It is widely accepted that, depending on the initiating circumstance, a varying proportion of the "dark" neurons recover in a few hours, especially in an otherwise hardly damaged environment, and if the post-injury circumstances are favorable (Csordás et al., 2003). Under unfavorable post-injury circumstances, the recovery may take a few days and the proportion of the dying "dark" neurons increases (Mázló et al., 2005). The death-mode of "dark" neurons is unanimously accepted to be necrosis, based on experiments in which dead "dark" neurons were removed from damaged (necrotic, excitotoxic or contused) brain tissues in the same way as the "true" necrotic cells; specifically, through swelling, dispersion and dissolution (Fig. 6a).

In contrast, by means of non-contusing head injury, condenser-discharge electric shock, mild hypoglycemia (Csordás et al., 2003; Gallyas et al., 2005) or transient focal ischemia (unpublished observation) we could produce "dark" neurons

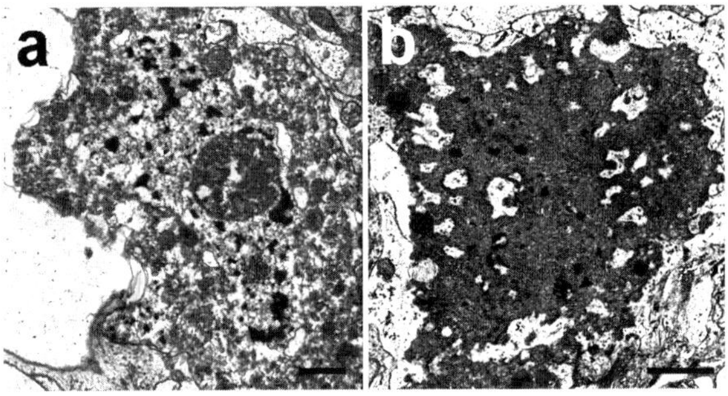

Fig. 6 Necrotic-like (**a**) and apoptotic-like (**b**) removal of "dark" neurons from a necrotic focus (**a**) and from its penumbra zone (**b**) in the caudo-putamen of a rat with transient focal cerebral ischemia. Scale bars: **a** and **b** = 1 μm

in apparently intact brain tissue. Several of them died. These were removed from the brain tissue in the same way as apoptotic neurons from intact environment (compare Fig. 6b with Fig 7a). Specifically, their cytoplasm underwent convolution and then fragmentation. The fragments remained compact, homogenous and membrane-bound, even after they had been engulfed by phagocytotic cells. These results suggest that the generally accepted relationship between "dark" neurons and necrosis is restricted to the capability of a seriously damaged environment to impose a necrotic-like disintegration upon the "dark" neurons that have died in some other way.

As regards the apoptotic neurons, after completion of the biochemical cascade and segregation of the nuclear chromatin, they are known to undergo a sudden dramatic cytoplasmic shrinkage. Their shrunken cytoplasm reveals the very same set of light- and electron-microscopic changes as that of "dark" neurons (Liposits et al., 1997; Gallyas et al., 2005). In addition, apoptotic neurons are also similar to "dark" neurons in the manner of their removal from a contused, an excitotoxic or a necrotic environment (compare Figs. 7b and 6a). These can impose a necrotic-like disintegration even upon apoptotic cells. Large clumps of aggregated nuclear chromatin are only indicative of the death type of these cells.

Bearing all these points in mind, we suggest here that the mechanism of the cytoplasmic compaction in apoptotic cells also involve the gel-to-gel phase transition in question. If this is true, than the formation of the cytoplasmic gel structure

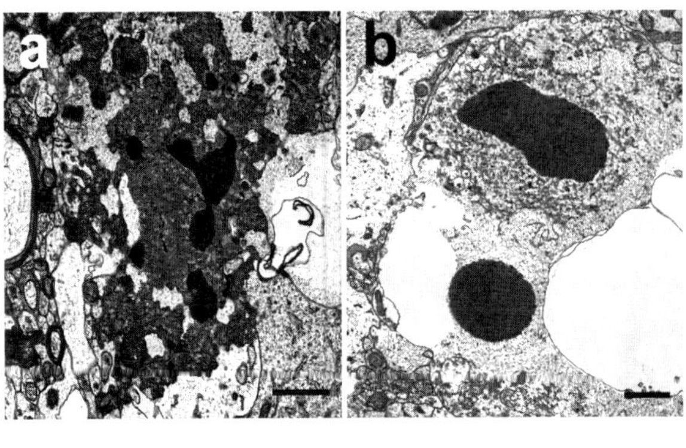

Fig. 7 Apoptotic (**a**) and necrotic-like (**b**) removal of apoptotic neurons from a necrotic focus (**b**) and from its penumbra zone (**a**) in the caudo-putamen of a rat having suffered transient focal cerebral ischemia. Scale bars: **a** and **b** = 1 μm

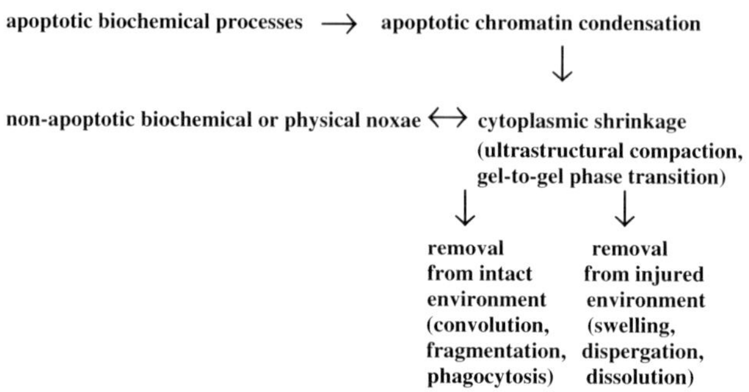

Fig. 8 Suggested relationship of the formation and fate of "dark" cells to the morphological execution of apoptosis

must be genetically programmed in neurons, and in many other cell types that also have a "dark" form (Harmon, 1987), for starting the morphological execution of apoptosis in the cytoplasm. If such a gel-to-gel phase transition is initiated by a non-apoptotic biochemical process, for instance by hypoglycemia or ischemia, or by a momentary mechanical or electric noxa, "dark" cells are produced, which may die via an unknown pathway, or may recover, depending on the exogenous circumstances. (Fig. 8). For this reason, the formation of "dark" cells might reflect a malfunction of the cytoplasmic gel structure, the physiological function of which is to take part in the morphological execution phase of ontogenetic apoptosis.

References

Cammermeyer, J., 1961, The importance of avoiding "dark" neurons in experimental neuropathology. Acta Neuropathol., 1:245–270.

Csordás, A., Mázló, M., and Gallyas, F., 2003, Recovery versus death of "dark" (compacted) neurons in non-impaired parenchymal environment. Light and electron microscopic observations. Acta Neuropathol., 106:37–49.

Gallyas, F., 1982, Physico-chemical mechanism of the argyrophil III reaction. Histochemistry, 74:409–421.

Gallyas, F., Güldner, F.H., Zoltay, G., and Wolff, J.R., 1990, Golgi-like demonstration of "dark" neurons with an argyrophil III method for experimental neuropathology. Acta Neuropathol., 79:620–628.

Gallyas, F., Farkas, O., and Mázló, M., 2002, Traumatic compaction of the axonal cytoskeleton induces argyrophilia: histological and theoretical importance. Acta Neuropathol., 103:36–42.

Gallyas, F., Farkas, O., and Mázló, M., 2004, Gel-to-gel phase transition may occur in mammalian cells: Mechanism of formation of "dark" (compacted) neurons. Biol. Cell, 96:313–324.

Gallyas, F., Csordás, A., Schwarz, A., and Mázló, M., 2005, "Dark" (compacted) neurons may not die through the necrotic pathway. Exp. Brain. Res., 160:473–486.

Harmon, B.V., 1987, An ultrastructural study of spontaneous cell death in mouse mastocytoma with particular reference to dark cells. J. Pathol., 153:345–355.

Hoffman, A.S., 1991, Conventional and environmental-sensitive hydrogels for medical and industrial uses: a review paper. Polymer Gels, 268:82–87.

Kellermayer, R., Zsombok, A., Auer, T., and Gallyas, F., 2006, Electric-shock experiments support the gel-phase transition theory of the formation of "dark" neurons. Cell Biol. Int., 30:175–182.

Kovács, B., Bukovics, P., and Gallyas, F., (2007), Morphological effects of transcardially perfused sodium dodecylsulfate on the rat brain: Cell-biologic aspects. Bio. Cell 99:425–432.

Kövesdi, E., Pál, J., and Gallyas, F., (2007), The fate of "dark" neurons produced by transient focal cerebral ischemia in non-necrotic and non-excitotoxic environment. Neurobiological aspects. Brain Res. 1147:472–483.

Ling, G.N., and Ochsenfeld, M.M., 1973, Mobility of potassium ion in frog muscle cells, both living and dead. Science, 181:78–81.

Ling, G.N., and Walton, C.L., 1975, What retains water in living cells. Science, 191:293–295.

Liposits, Zs., Kalló, I., Hrabovszky, E., and Gallyas, F. , 1997, Ultrastructural pathology of degenerating "dark" granule cells in the hippocampal dentate gyrus of adrenalectomized rats. Acta Biol. Hung., 48:173–187.

Mázló, M., Gasz, B., Szigeti, A., Zsombok, A., and Gallyas, F., 2005, Debris of "dark" (compacted) neurones are eliminated from an otherwise undamaged environment mainly by astrocytes via blood vessels. J. Neurocytol., 23:559–569.

Newman, G.R., and Jasani, A.B., 1998, Silver development in microscopy and bioanalysis: past and present. Review article. J. Pathol., 186:119–125.

Pollack, G.H., 2001, Cells, Gels and the Engines of Life. Ebner and Sons, Seattle, pp. 1–298.

Pollack, G.H., 2002, The cell as a biomaterial. J. Mater. Sci. Mater. Med., 13:811–831.

Tanaka, T., Sato, E., Hirokawa, Y., Hirotsu, S., and Peetermans, J., 1985, Critical kinetics of volume phase transition of gels. Phys. Rev. Lett., 55:2455–2458.

Tanaka, T., Annaka, M., Ilmain, F., Ishii, K., Kokufuta, E., Suzuki, A., and Tokita, M., 1992, Phase transitions of gels. In: Karalis, T.K. (Ed.), Mechanics of Swelling. Springer, Berlin, NATO ASI Series, Vol. H64, pp. 683–703.

Zsombok, A., Tóth, Zs., and Gallyas, F., 2005, Basophilia, acidophilia and argyrophilia of "dark" (compacted) neurons in the course of their formation, recovery or death in an otherwise undamaged environment. J. Neurosci. Meth., 142:145–152.

Puzzles of Cell and Animal Physiology in View of the Chain-Ordering Transition in Lipid Membrane

D.P. Kharakoz

Abstract The chapter discusses the concept of a membrane phase-transition-dependent (MPTD) elementary machine, implementing physiological functions in the cell. Synaptic exocytosis is among the most important functions, in which the highest perfection is achieved owing to the MPTD machine. The concept presents a unitary approach to different aspects of physiology and allows to answer a number of general unsolved questions, which have been puzzling investigators for decades. The following questions, among many, have been offered a reasonable answer: What is the biological expedience of homeothermy? Why is anaesthesia reversed by hydrostatic pressure, while the pressure itself causes immobilisation? What is the recovery function of sleep, and why is this function incompatible with the active state of warm-blooded animals?

Keywords Biological membranes · lipids · chain-ordering phase transition · synaptic exocytosis · thermoregulation · homeothermy · general anaesthesia · pressure-reversal of anaesthesia · high-pressure neurological syndrome · function of sleep · calcium · molecular machines

1 Introduction

A large body of observations indicate that the phase state of biological membranes is physiologically important. This is not only because the standard state of the membranes is a fluid one; of great importance is also the transition from fluid to solid state called main or chain-ordering transition (Melchior and Steim, 1976; Shnoll, 1979; Biltonen, 1990; Bloom et al., 1991; Hazel, 1995; Kharakoz, 2001; Mouritsen, 2005).

In the 1970s, when the protein-based mechanisms of cell machinery were dominating, Simon E. Shnoll was the first who explicitly formulated the idea that phase transition in lipid membrane certainly plays a key functional role in the mechanisms

D.P. Kharakoz
Institute of Theoretical and Experimental Biophysics, Russian Academy of Sciences, Institutskaya street 3, 142290 Pushchino, Russia

of the cell and brain excitability (Shnoll, 1979). This conclusion resulted from three schools of thoughts that had been merged in his book.

The first one originated from Ernest Overton's theory of the cell wall assumed to be of lipoid nature (Overton, 1901) – the theory further developed into a central concept of cell biology, the concept of semipermeable cellular membranes in which lipid bilayer serves as the basic structural component.

The second school was initiated by Russian cytologist Dmitry N. Nasonov, who developed ideas that the cell excitability is based on nonspecific reactions of the protoplasm considered as a colloidal system able to assume different phase states (Nasonov and Alexandrov, 1949). His ideas became known to Western scientists through English-translated books of his follower, Afanasy S. Troshin (Troshin, 1956, 1966). This line of thinking continued in a brilliant and provocative book by Gerald Pollack, (Pollack, 2001).

The last flow of ideas had been originated from the famous school of experimental biologists created in the first half of 20th century by Nikolay K. Koltsoff, the scientist who, reflecting on the physicochemical grounds of heredity has formulated the central biological idea of 20th century, the concept of matrix replication of biological polymers carrying genetic information (Koltsoff, 1927; Haldane, 1945). His follower, Nikolay W. Timofeeff-Ressoffsky has elaborated, in discussions with Max Delbrück and Paul Dirac in early 1930s, the evolutionary principle of convariant replication (see Mednikov, 2001, and a biographical essay in the book by Shnoll, 2001).

The latter principle was further added by Shnoll (1979) with the principle of maximal (physically possible) kinetic perfection achieved by any functional system at each level of evolutionary development. Evolutionary analysis of physiological and physicochemical observations led Shnoll to the conclusion that it is brain whose excitability must crucially depend on the melting/ordering phase transition in lipid membrane. He thought it is the nerve pulse propagation which should be a transition-dependent process.

The latter idea, independently elaborated in later works (Kinnunen and Virtanen, 1986; Heimburg and Jackson, 2005) needs a special discussion, which however is beyond the scope of this chapter. Instead, I shall focus on another key point of nerve excitability. This is synaptic transmission, a process that, in my opinion, could not be fast enough should it not be based on phase transition in the presynaptic membrane (Kharakoz, 2001). I will begin from formulation of a hypothetical *membrane phase-transition-dependent (MPTD) molecular machine*, which is postulated to control some physiological functions. Then the expected physiological consequences of the hypothesis will be considered and compared with the physiological observations, some of which have been puzzling investigators for many decades.

2 Membrane Phase-Transition-Dependent Machine (MPTD Machine) and its Operating Characteristics

Analysing processes at the level of macromolecules and molecular ensembles, one should recall a general biophysical idea that in the molecular "machines", the

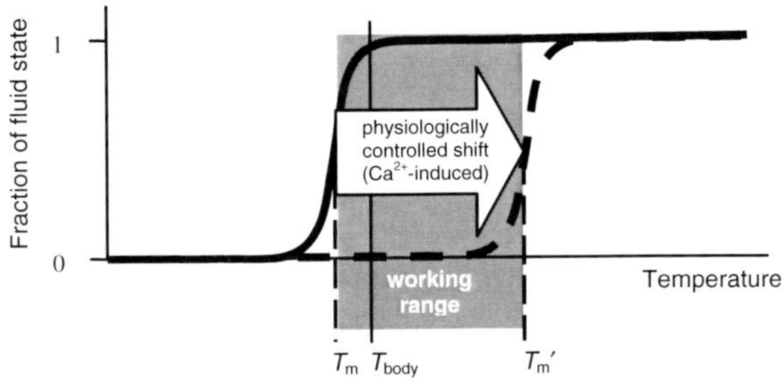

Fig. 1 The working range of an elementary membrane phase-transition-dependent machine. Here T_m (at the solid curve) is the transition temperature between the gel and liquid-crystalline states in the absence of physiological effectors and T'_m (at the *dashed* curve) is the transition temperature shifted upwards by a physiological effector, e.g., by Ca^{2+} ions. The working range of temperatures (marked with grey halftone) is therefore defined as the range of body temperatures (T_{body}) where the membrane can be reversibly switched by the effector between the fluid and solid states

fastest large-scale structural rearrangements are possible when they are essentially mechanical processes (Blumenfeld, 1981). The concept discussed here and in the following sections is developed in line with this way of thinking.

Let us consider a hypothetical elementary membrane phase-transition-dependent (MPTD) molecular machine. Its basic part is a biological membrane (or a membrane domain) which undergoes, under physiological control, the phase transition between the liquid-crystalline and gel states. Let us also assume that there are physiological functions modulated by these alternating states.

How can phase state be alternated under physiological control? Figure 1 shows qualitatively the dependence of phase state of a lipid membrane on temperature. Various physiologically relevant parameters can induce a shift of the transition temperature, T_m (reviewed by Kharakoz, 2001). The transition between the phase states can therefore be induced by physiological effectors. For instance, Ca^{2+} ions can increase T_m due to their interaction with the negative net charge of biological membranes. What is important is that the Ca^{2+} concentration can change very fast: in synaptic terminals, for instance, it takes fractions of millisecond (Augustine et al., 1987). Let us assume calcium ions to be the major and fastest physiological effector of phase state. The curves presented in Fig. 1 can serve as the main operating characteristic of the elementary MPTD machine. Let us consider its basic properties.

3 Where in the Cell Might the MPTD Machine Work?

The phase state of membrane can participate in physiological processes both indirectly and directly. An indirect function can be realized through mechanical lateral and bending stresses modulating the activity of membrane proteins. Examples of such protein-mediated functions have been comprehensively reviewed in the book by Mouritsen (2005); among them is a possibility of mechanical control of ion channels. Direct functions of phase transition can be implemented through the transition-induced membrane curvature, lateral stress, pore formation and rupture of the membrane (Antonov et al., 1980, 2007; Cevc and Richardsen, 1999). Among the most important properties of the chain-ordering transition is a dramatically reduced surface area of the membrane, by about 20% (Laggner et al., 1987; Lohner et al., 1987; Heimburg, 1998). Possible physiological functions resulting from this property are considered below.

3.1 Cell Shaping

As mentioned, the surface area is dramatically reduced upon the fluid-to-solid transition. This means that an asymmetrically provoked transition, when only one monolayer solidifies and the other remains unchanged, will cause a stress which would result in a certain change of bilayer curvature (Kharakoz, 2001; Mouritsen, 2005). Therefore, formation of cell shape (during cell division, e.g.) could be controlled not only by the proteins of cytoskeleton but also by asymmetric phase transitions in the membrane.

3.2 Facilitation of Membrane Fusion

The same mechanical behaviour of an asymmetrically solidified membrane might also play an important role in the formation of intermediate states in membrane fusion because bilayers are bent in the intermediate states (Cevc and Richardsen, 1999). An interesting situation could arise if asymmetric solidification is induced in a vesicle. Then contraction of the solid monolayer would cause its rupture, resulting in a transient exposure of the nonpolar membrane interior which would facilitate fusion. This is another possibility of direct involvement of phase transition in the fusion process.

3.3 Transient Water Pores and Ionic Permeability

Phase transition can cause a temporary membrane rupture and formation of a transient water pore. A dramatically increased ionic permeability of lipid membranes at phase transition is a well documented fact (Papahadjopoulos et al., 1973; Antonov et al., 1980, 2007).

3.4 Nerve Pulse Generation and Propagation

As mentioned in Introduction, Shnoll has suggested that the mechanism of nerve pulse propagation should depend on the membrane phase transition somewhere in the brain. This idea formulated by Shnoll in general terms is getting now a more distinct shape. As follows from the preceding paragraph, ionic permeability is controlled not only via the conformation state of channel proteins but also directly, by modulation of the lipid phase state. From this fact, Kinnunen and Virtanen (1986) proposed a qualitative mechanism of nerve pulse propagation depending on the physiologically controlled phase transition in biological membrane. The idea of the phase-transition-dependent action potential has recently been revisited by Heimburg and Jackson (2005). They suggested a theoretical model, in which the non-linear mechanical properties of membrane near phase transition provide conditions for a mechanical soliton propagation along the axon. Some problems of the classical Hodgkin-Huxley theory of action potential could find a solution within these hypotheses.

3.5 Phase-Transition-Dependent Exocytosis

Exocytosis is a basic process in eukaryotes. A special and very important case of this process, the synaptic exocytosis, was considered several years ago within the framework of the hypothesis that in fast synapses exocytosis is driven by the membrane phase transition (Kharakoz, 2001). A qualitative model of the process has been suggested (illustrated in Fig. 2). It is based on the widely accepted knowledge about the mechanism of exocytosis in central synapses, with only one essential modification:

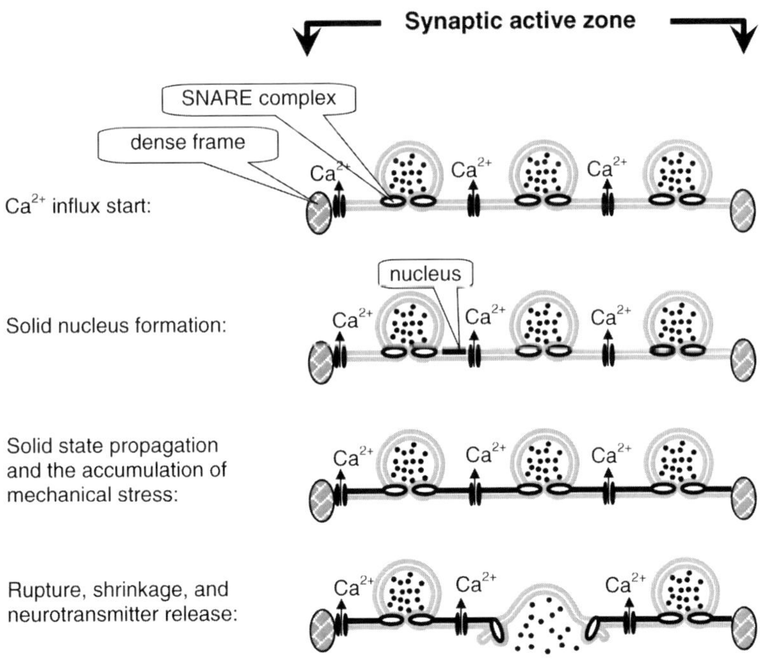

Fig. 2 Fast synaptic exocytosis resulting from the action of an MPTD machine. The sequence of steps is shown from top to bottom. The diagram is drawn according to the model developed earlier (Kharakoz, 2001)

the final step of the process, when a docked and primed synaptic vesicle responds to the peak of Ca^{2+} concentration, is assumed to be driven by the Ca^{2+}-induced phase transition in the presynaptic membrane of the active synaptic zone.

4 Basic Requirements and Functional Properties of the MPTD Machine

In order to understand the physiological manifestations of the MPTD machine we should explore what are the expected functional properties of the machine. Considered here will be a Ca^{2+}-controlled machine (Fig. 1).

4.1 The Main Requirement to the Molecular Composition: A Physicochemical Compound

The sharpest phase transition (occurring within the narrowest temperature range) is observed if the system is homogeneous and both phases have the same molecular composition. The simplest examples of such behaviour are represented by pure *one*-component systems (Lee, 1977), which certainly cannot be found among biological membranes. The other possibility is realized if a *multi*-component system is essentially a physicochemical compound. Compound is a mixture with a certain stoichiometric ratio of individual components that corresponds to the stoichiometry of a perfectly packed crystal. Phase behaviour displayed by such mixtures is very similar to that of one-component systems: their phase transition from fluid to solid state is sharp. This phenomenon, called compound formation, is well-known in the physical chemistry of alloys (Dreving and Kalashnikov, 1964). The cases of such behaviour were also reported for lipid systems (Iwahashi et al., 2001; Lopez-Garcia et al., 1994).

Therefore, in order to display a sharp transition, the molecular composition of the membrane in MPTD machine should be a physicochemical compound.

4.2 The Width of Working Range of Temperatures

This width is determined by two factors: the sensitivity of phase transition temperature (T_m) to Ca^{2+} ions and the maximal possible concentration of free Ca^{2+} ions in the cell. The sensitivity depends mainly on the net electric charge on membrane surface. The earlier estimation (Kharakoz, 2001) made on the basis of experimental data reported for model lipid systems showed that a membrane, containing 10 mole% of a negatively charged lipid, had a sensitivity, $\Delta T_m/\Delta[Ca^{2+}]$, of 3–10 K/mM. As for the maximal concentration of Ca^{2+}, it certainly cannot exceed that in the extracellular space, which is about 1 mM. In synaptic terminals, the transient peak of $[Ca^{2+}]$ is about 0.3 mM and, hence, *the maximal shift of T_m in this case is* *of the order of 1–3 K.* This value can be considered as a rough estimate for the width of working range (see Fig. 1).

4.3 Volume Change and the Sensitivity to Bulk Pressure

Chain melting transition in lipid membranes is accompanied by a volume increase ($\Delta V > 0$). Therefore, hydrostatic pressure causes T_m to increase (according to Clapeyron-Clausius equation: $dT_m/dP = T_m \Delta V/\Delta H$, where ΔH is the transition enthalpy). Experimental data show that for both the biological and model membranes, the sensitivity of T_m to pressure varies within $\Delta T_m/\Delta P = 0.015 - 0.030$ K/atm (Macdonald, 1984). An important conclusion from the sensitivity being in this range is that the hydrostatic pressure of 100 atm will increase working range of temperatures by a value comparable with the width of the range (1.5–3 K). Therefore, this high pressure is able to "switch off" the MPTD machine if the shift is not compensated by the corresponding increase of body temperature. Physiological consequences of this fact are discussed below.

4.4 Surface Area Change and the Sensitivity to Lateral Pressure

Surface area also dramatically increases during melting transition, by 25% (Laggner et al., 1987; Lohner et al., 1987; Heimburg, 1998). Therefore, lateral pressure also causes working temperatures to increase. This is one of the basic points in the soliton-like mechanism proposed recently to participate in the process of the nerve pulse propagation (Heimburg and Jackson, 2005).

4.5 Influence of Anaesthetics

A common physicochemical property of all general anaesthetics is their ability to decrease the melting transition temperature in lipid membranes (Kharakoz, 2001, and refs. therein). Therefore, by adding anaesthetics to the system, one will move working range of the MPTD machine towards lower temperatures (Fig. 1). At a certain anaesthetic dose, the critical shift is achieved: the body temperature gets out of the working range and the machine stops (if the body temperature is not decreased in order to compensate the anaesthetic action). The critical shift should be of order of the working range width – as it really is, as discussed below. Different anaesthetic drugs have different ability to decrease T_m, and this ability is well correlated with the anaesthetic potency of drugs. This correlation is discussed in next section. (A particular case of the influence of anaesthetics on the phase-transition-dependent mechanism of soliton propagation has been considered recently by Heimburg and Jackson, 2007).

4.6 Influence of Natural Metabolites and the Temperature Adaptation of Phase State

In principle, natural metabolites may change the transition temperature in biomembranes. Unfortunately, little is known in this field. Some data are available on the saturated fatty acids: under certain conditions, their interaction with lipid membranes may increase T_m (Agafonov et al., 2003). It is also well known that the transition temperature depends on the lipid composition (see, e.g., Hochachka and Somero, 1984; Hazel, 1995; Hazel et al., 1998). It is not inconceivable that the endogenously produced ethanol may take part in the temperature adaptation reactions in goldfish (Crawshaw et al., 1989).

4.7 Kinetics of the MPTD Machine and its Dependence on the Lipid Chain Length

Considering the kinetics of the Ca^{2+}-induced phase transition, one can subdivide the whole process into distinct steps. The first step is the initiation, during which Ca^{2+} concentration is increasing until the moment when T_m exceeds the body temperature (T_b). The system thus gets into a metastable state in which the membrane, being still fluid, is prepared to "jump" into the solid state. Further events, until the whole membrane gets solid, may be considered as a second step – the step of emergence of a new phase. It includes the formation of a nucleus of solid state and the propagation of the state all over the membrane. It would be also rational to distinguish a third step, which would include all the events related to the transformation of the membrane solidification into the ultimate physiological function. In case of the MPTD exocytosis described above, this will be the transformation of mechanical stress into the membrane fusion and rupture. Let us consider the factors that dominate in determining the kinetics of the whole process.

The time of the first, initiation step (τ_I) depends on how close is the transition temperature to the temperature of the body (Fig. 3). The less the difference T_{body} – T_m, the faster will be the process of reaching the metastable state (given that all other conditions, including Ca^{2+} influx rate and Ca^{2+}-sensitivity of membrane, are equal). However, this difference should not be less than the transition half-width δ, otherwise a spontaneous solidification would occur due to fluctuations, and the cell would lose the control over the process. Therefore, in order to keep T_m closer to T_{body} and thus reduce the initiation time, the half-width δ should be decreased as much as possible (because $\tau_I \sim \delta$). On the other hand, among the dominant structural parameters, determining the transition half-width, is the chain length (n): the longer are the lipid chains, the sharper is the transition (Uchida et al., 1997). This dependence roughly follows the law: $\delta \sim n^{-6}$ (Kharakoz, unpublished). Therefore, the longer are the chains of the lipid the MPTD machine is made of, the shorter is the initiation time, which will follow the law of $\tau_I \sim n^{-6}$.

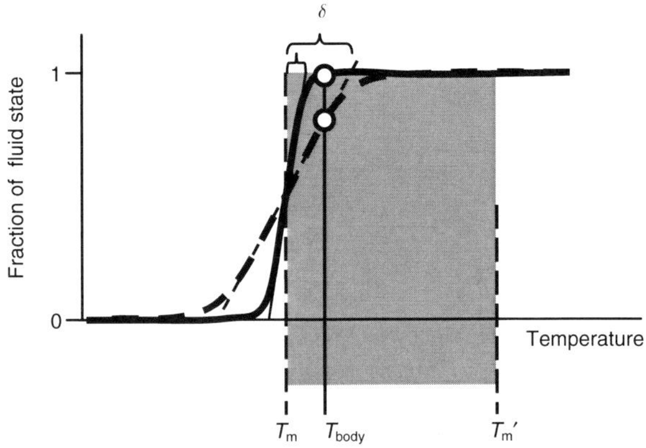

Fig. 3 Comparison of the transition curves that differ in sharpness. The sharper the transition (*solid line* compared to *dashed* one), the closer will be T_{body} to T_m – without loss of the control of phase state. The half-width of transition is designated by δ; other designations are the same as in Fig. 1

The rate of the second step (phase development) is restricted by the probability of nucleus formation, which is a function of energy of the solid/fluid interface: as the energy increases, the probability sharply decreases, because the nucleus becomes more energetically expensive (Ubbelohde, 1965; Kharakoz, 2003). This interfacial energy, in turn, is larger in the longer lipids (Kharakoz et al., 2006). Therefore, the nucleus probability and, hence, the rate of phase development *would diminish as the lipid chain length increases*.

The third step will depend on the nature of the physiological function considered. Let us consider the model of phase-transition-driven synaptic exocytosis (Section 3.5), which is of particular interest to us. In the case of this model, the third step is the fusion of the vesicular and plasma membranes. This process is conjugated with the membrane bending; and the thicker the membrane, the more energy it would require to bend. Therefore, *fusion should be delayed in the membranes made of longer lipids*. The delay should be a sharp non-linear function of the length.

Summarizing these considerations illustrated in Fig. 4, one comes to the conclusion that there should exist an optimal chain length, at which the MPTD machine would work at the fastest possible rate. The machine cannot function fast when the lipid chains are short, because rate gets limited at the stage of transition initiation. It also cannot work fast if the chains are long; in that case, it is the solid phase development that becomes a bottle-neck, limiting the rate of processes. Taking into account the well-known positive correlation between the phase transition temperature and the chain length (Cevc and Marsh, 1987), we come to an important conclusion that *there should be an optimal range of temperatures where the MPTD machine will work at the highest possible rate*. This conclusion will be recalled in Section 5.6 where a puzzle of the evolution of warm-blooded will be considered.

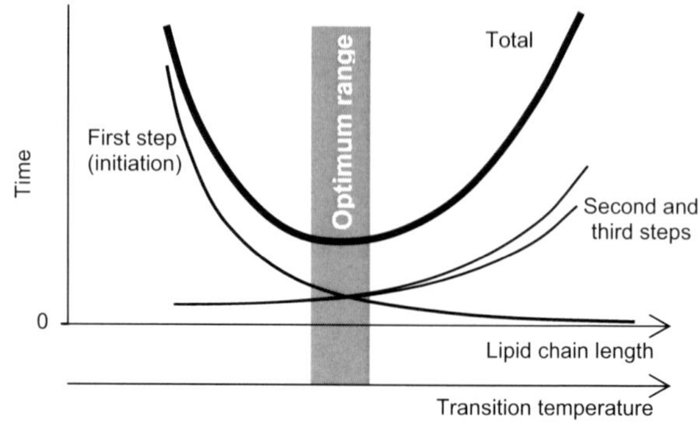

Fig. 4 Total time of transition is a sum of contribution of steps. The picture illustrates the origin of an optimum range of chain length (and temperature) where the transition proceeds with the shortest possible time (marked with grey half-tone)

4.8 A Problem of Self-Maintenance of Membrane Compound

This problem can be formulated in the style of a postulate: *If biological evolution has found a membrane compound that can be used as a working medium for the membrane phase-transition-dependent machine then it should "invent" a system for self-maintenance (purification) of the compound.* The mechanism of purification is strongly required, because a compound surrounded with the areas of a different molecular composition will inevitably accumulate impurities, and, hence, the necessary sharpness of phase transition will be lost and the function will be impaired. No specific mechanism is known for recovering the exact composition and structure of the functional lipid domains. At the same time, a *non-specific* way of purification is well-known – this is the "crystallization" from melt. This seems to be the only possibility of self-maintenance of a membrane compound. Note that to initiate a purification of this kind, the body temperature must be set below the solidification point. Therefore, the MPTD machine has to be inevitably switched off for the period of "re-crystallization". This conclusion will help us to understand the puzzles of sleep considered below.

4.9 Hierarchy and Interplay of Times in the MPTD Machine

Among the factors, influencing phase transition temperature, there are fast and slow variables. The fastest one is the Ca^{2+} concentration that can rise to its effective level for a very short time. In excited synaptic terminals, this will take fractions of a millisecond (Augustine et al., 1987). On the other hand, some variables are slow. For instance, adaptive change of lipid composition in cellular membranes may take days. The metabolic processes that influence membrane fluidity and transition temperature can have any intermediate characteristic time: it is determined by rates of the corresponding enzymatic processes, such as the reactions of lipid saturation/desaturation (catalysed by saturases) or lipid hydrolysis (catalysed by lipases). For review of the membrane adaptation processes, see the books by Alexandrov (1977) and Hochachka and Somero (1984). Therefore, the variables that influence the phase state of the MPTD machine show a hierarchy of times, from fractions of a millisecond to days, with the functional properties of the machine being determined by interplay between them. One can assume that the slow variables control the adaptation processes, while the fast variables are used in the cell as continually operating variables that trigger the state of the machine when its physiological function is implemented.

5 Biological Observations in View of the MPTD Machine Concept

5.1 *Puzzles in the Kinetics of Exocytosis*

Exocytosis in the central synapses is characterized by two unique features (Goda and Südhof, 1997), which surprise investigators. Despite an active zone contains a dozen of vesicles that are fully prepared to release their content, most nerve impulses, reaching the synapse, do not elicit exocytosis. In hippocampus, less than a third of all action potentials results in the neurotransmitter release, indicating exocytosis to be essentially a stochastic process. Moreover, when the release still occurs, never more than one vesicle fuses with the presynaptic membrane, indicating the existence of a "lateral inhibition" mechanism: a singe fusion event becomes immediately "known" to other vesicles within the same active zone. How can the inhibitory signal be transmitted over the active zone in such a short time?

Both these features find a simple explanation within the phase-transition-dependent mechanism. The stochastic character of fusion is a direct consequence of the stochastic nature of solid-state nucleation (for lipids, this problem was briefly considered by Kharakoz, 2001). If a nucleus does not form during the calcium peak, fusion does not occur. If the nucleus is formed and starts to propagate over the plasma membrane, the membrane will grow stressed until the first rupture-and-fusion event. The incorporation of vesicle's lipid into the plasma membrane immediately takes the stress away; thus, no more vesicles can fuse in the same zone until the initial state of plasma membrane is recovered (which will take a longer time). Therefore, the inhibitory signal propagates mechanically, with a speed typical for mechanical waves.

Whether or not one fusion event is enough to release the mechanical stress caused by the solid monolayer contraction? This question, important in the context of the mechanical inhibitory effect, can be answered within the following geometric consideration. Characteristic size of active zones is of order of 200 nm (Moshkov, 1985); this is the diameter in the case of compact zones approximated by circles or the width in the case of extended zones. Let us assume, for simplicity, that the active zone is circle-shaped. Then, its surface area is roughly 30×10^3 nm^2. As mentioned, solidification of the membrane induces 20% contraction of its area, which amounts to 6×10^3 nm^2. This is the area deficiency to be compensated by the material of a vesicle incorporated upon the exocytosis event. Diameter of typical synaptic vesicle, 50 nm, corresponds to surface area

8×10^3 nm^2. We see this is enough to fully compensate the solidification-induced stress.

5.2 Pressure-Induced Hyperthermia

The temperature working range of the MPTD machine increases with pressure (Section 4.3). One might therefore expect that a "discomfort" experienced by animals under extra pressure, should initiate their thermoregulatory reaction aimed to increase their body temperature. This expectation was justified by the study of behavioural thermoregulation in a small aquatic animal, euryhaline gammarid *Parhyale hawaiiensis* (Kinney et al., 1981). Figure 5 shows that the preferred temperature does increase along with the elevated pressure. The increase is concordant with the slope observed for the phase transition in model membranes.

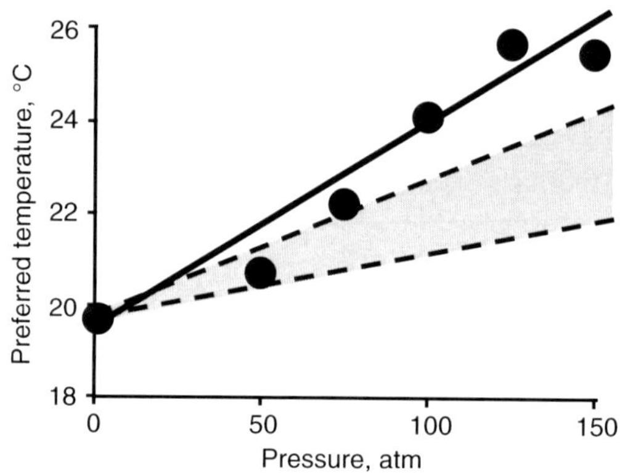

Fig. 5 Preferred temperature of euryhaline gammarid Parhyale hawaiiensis as a function of hydrostatic pressure (experimental data were taken from Kinney et al., 1981). *Solid line* shows the average slope of the experimental dependence: $\Delta T_{\text{body}}/\Delta P = 0.045K/\text{atm}$. A half-tone sector between the two *dashed lines* shows the range of experimental data for the pressure dependence of transition temperature in model membranes composed of synthetic or natural lipids ($\Delta T_{\text{m}}/\Delta P = 0.015 - 0.03K/\text{atm}$; Macdonald, 1984)

5.3 Anaesthetic Hypothermia

Anaesthetics decrease transition temperature in model membranes and thus move the working range towards lower temperatures (Section 4.5). It is expected that the body temperature should also decrease to compensate for the action of anaesthetics. Indeed, the anaesthetic-induced hypothermia in animals is a well-known phenomenon. A collection of data was analysed earlier (Kharakoz, 2001). Here a summary picture is presented (Fig. 6), where the anaesthetic potency of substances

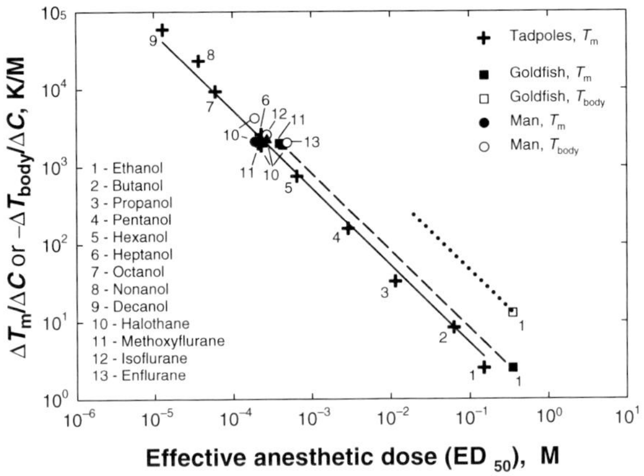

Fig. 6 Correlation between the anaesthetic potency of solutes (ED_{50}) and their potencies to decrease transition temperature in model membrane ($-\Delta T_m/\Delta C$, crosses and filled symbols) and to cause hypothermia in animals ($-\Delta T_{body}/\Delta C$, open symbols). The straight lines are drawn according to the law $ED_{50} \times \Delta T/\Delta C = a$ for the following values of the constant a: -0.53 K (*solid line*), -0.83 K (*dashed line*), and -3.7 K (*dotted line*). The figure is adapted from Kharakoz, 2001

is plotted vs. their potency to cause hypothermia ($-\Delta T_{body}/\Delta C$), and their potency to lower T_m in simple model membranes ($-\Delta T_m/\Delta C$). In the figure, C is the molar concentration of anaesthetic in water or in animal's body and T_{body} is the temperature measured either in water environment (in case of small aquatic animals) or in the brain (in case of other animals). Anaesthetic potency is expressed in terms of the effective dose, ED_{50}, a concentration of anaesthetic that causes anaesthesia in 50% animals. A good correlation between the anaesthetic potency and the ability to decrease T_m is observed.

Despite a great difference in ED_{50} (of several orders of magnitude), the substances tested lower T_m by invariant value, 0.7 °C per an effective anaesthetic dose. Of more interest from the viewpoint of the MPTD machine concept is the hypothermic effect in animals. Small aquatic animals are the most appropriate objects for such analysis because temperature of their body and brain can be easily controlled in the experiment. As follows from the data reported by O'Connor et al. (1988) on the behavioural thermoregulation in goldfish, the hypothermic potency of ethanol ($\Delta T_{body}/\Delta C$) is -13°C/M, and its anaesthetic dose (ED_{50}) is 0.35 M. Therefore, the hypothermic effect of one ED_{50} ($\Delta_{anesth} T_{m,0}$) is -3.7°C. This value can be considered as an estimate of *the critical shift of transition temperature in the synaptic membrane required to achieve anaesthesia.* The meaning of this quantity is clarified in Fig. 1 and in Section 4.2: this is a measure of width of the MPTD machine working range. One can see that determined more directly from the physiological experimental data, this value is close to that roughly estimated in Section 4.2.

5.4 Shortcomings of the Meyer-Overton Rule and a Better Alternative

Known more than 100 years is the empirical Meyer-Overton rule, which says that the anaesthetic potency of a substance correlates directly with its oil/water partition ratio (Meyer, 1899; Overton, 1901; Ueda, 2001). There are, however, systematic deviations from this rule. For instance, in the homological series of saturated normal-chain alcohols the anaesthetic potency increases together with the partition ratio until the chain length reaches 12 C-atoms (Kaminoh et al., 1992); the longer molecules continue to show increasing solubility in oil but abruptly lose the anaesthetic potency. This phenomenon is known as the "cut-off effect". It was also noticed that just these long-chain alcohols lost their ability to decrease the main transition temperature in model membranes. As follows from the preceding subsection, this correlation has a clear relevance to the mechanism of anaesthesia emerged from the MPTD machine concept, because it states that the ability of a substance to decrease working temperature of the machine is *directly* related to its ability to switch the machine off immediately after the temperature shift has reached a critical value.

5.5 The Puzzle of Pressure-Reversal of Anaesthesia

Pressure-reversal of anaesthesia is a well-known general phenomenon. The essence of the phenomenon is that the immobilization caused by anaesthetics is reversed by a high pressure. How can one explain this – in view of the fact that applied in the absence of anaesthetics, high pressure (~100 atm) itself causes immobilization, which is termed "high pressure neurological syndrome" (Dluzewski et al., 1983; Smith et al., 1984)? This observation illustrated in Fig. 6 is in accord with the properties of the MPTD machine, because anaesthetics and pressure have opposite effects on the transition temperature (Section 4.3 and 4.5), and the working range lowered by anaesthetics will hence be shifted back by the increased pressure.

Note that the high-pressure neurological syndrome is induced by the same high pressures (~100 atm), as those that cause the critical shift of working range of temperature in the MPTD machine, i.e. the shift compared with the working range width, ~3 K (cf. Section 4.3).

A qualitative similarity between the pressure-reversal of anaesthesia and the pressure-reversal of the anaesthetic-induced decrease of T_m in model membranes was noticed long ago (Trudell et al., 1975; Kamaya et al., 1979). However, it has been discussed in quite general terms, in the light of the idea that the central role in cell physiology belongs to proteins. According to this idea, the only way to explain the phenomenon is to say that the functional properties of membrane proteins *somehow* depend on the membrane phase state. The concept of the MPTD exocytosis in fast synapses (Section 3) suggests another, more direct and reasonable, explanation.

Cevc, G., and Marsh, D. 1987. *Phospholipid Bilayers. Physical Principles and Models.* Wiley, New York.

Cevc, G., and Richardsen, H. 1999. Lipid vesicles and membrane fusion. *Adv. Drug Deliv. Rev.* **38**:207–232.

Crawshaw, L.I., Wollmuth, L.P., and O'Connor, C.S. 1989. Intracranial ethanol and ambient anoxia elicit selection of coller water by gold fish. *Am. J. Physiol.* **256**:R133–R137.

Dluzewski, A.R., Halsey, M.J., and Simmonds, A.C. 1983. Membrane interactions with general and local anaesthetics: a review of molecular hypotheses of anaesthesia. *Mol. Aspects. Med.* **6**:461–573.

Dreving, V.P., and Kalashnikov, Ya.A. 1964. [*Phase Rule*]. Moscow University Publisher, Moscow. (In Russian)

Fuller, A., Moss, D.G., Skinner, J.D., Jessen, P.T., Mitchell, G., and Mitchell, D. (1999) Brain, abdominal and arterial blood temperatures of free-ranging eland in their natural habitat. *Pflugers Arch.* **438**:671–680.

Gillis, D.M. 1996. Why sleep? *Bioscience* **46**:391–393.

Goda, Y., and Südhof, T.C. 1997. Calcium regulation of neurotransmitter release: reliably unreliable? *Curr. Opin. Cell Biol.* **9**:513–518.

Haldane, J.B.C. 1945. A physisist looks at genetics. *Nature* **155**:375–376.

Hazel, J.R. 1995. Thermal adaptation in biological membranes: is homeoviscous adaptation the explanation? *Annu. Rev. Physiol.* **57**:19–42.

Hazel, J.R., McKinley, S.J., and Gerrits, M.F. 1998. Thermal acclimation of phase behavior in plasma membrane lipids of rainbow trout hepatocytes. *Am. J. Physiol.* **275**:R861–R869.

Heimburg T. 1998. Mechanical aspects of membrane thermodynamics. Estimation of the mechanical properties of lipid membranes close to the chain melting transition from calorimetry. *Biochim. Biophys. Acta* **1415**:147–162.

Heimburg, T., and Jackson, A.D. 2005. On soliton propagation in biomembranes and nerves. *Proc. Natl. Acad. Sci. USA* **102**:9790–9795

Heimburg, T., and Jackson, A.D. 2007. The thermodynamics of general anesthesia. *Biophys J.* **106**:099754.

Hochachka, P.W., and Somero, G.N. 1984. *Biochemical Adaptation.* Princeton University Press, Princeton, (Chap 11).

Ivanov, K.P. 1972. *Bioenergetics and Temperature Homeostasis.* Nauka, Moscow. (In Russian)

Ivanov, K.P., and Slepchuk, N.A. 1985. Sensitivity and precision of the operation of the human physiological thermostat. *Dokl. Akad. Nauk SSSR* **281**:753–757. (In Russian)

Iwahashi, M., Minami, H., Suzuki, T., Koyanagi, M., Yao, H., Ema, K., and Ashizawa, K. 2001. Thermotropic properties of steroids: 5-cholesten-3β-ol, 5α-cholestan-3β-ol, and 5β-cholestan-3α-ol. *J. Oleo. Sci.* **50**:693–695.

Kamaya, H., Ueda, I., Moore, P.S., and Eyring, H. 1979. Antagonism between high pressure and anesthetics in the thermal phase-transition of dipalmitoyl phosphatidylcholine bilayer. *Biochim. Biophys. Acta* **550**:131–137.

Kaminoh, Y., Nishimura, S., Kamaya, H., and Ueda, I. 1992. Alcohol interaction with high entropy states of macromolecules: critical temperature hypothesis for anesthesia cutoff. *Biochim. Biophys. Acta* **1106**:335–343.

Kharakoz, D.P. 2000. Phase transition in lipids and the problem of homeothermy. *Biophysics*, **45**(3):554–557. (In Russian)

Kharakoz, D.P. 2001. Phase-transition-driven synaptic exocytosis: A hypothesis and its physiological and evolutionary implications. *Biosci. Rep.* **21**(6):801–830.

Kharakoz, D.P. 2003. Chain-ordering phase transition in bilayer: Kinetic mechanism and its physicochemical and physiological implications. In: *Planar Lipid Bilayers and Their Applications.* (H.T. Tien and A. Ottova, eds.) Elsevier, Amsterdam-New York, pp. 239–268 (Chap 7).

Kharakoz, D.P., Panchelyuga, M.S., Tiktopulo, E.I., and Shlyapnikova, E.A. 2006. Gel-state nucleation in multilamellar vesicles of dimyristoylphosphatidylcholine and its relation to the critical temperature: A lattice model and microcalorimetry. *Web-archive of Cornell: Condensed Matter/0601523.* http://arxiv.org/abs/cond-mat/0601523.

Kinney, M., Jones, W.R., Royal, R., Brauer, R.W., and Sorrel, F.Y. 1981. A gradient tube system for the study of the effect of high hydrostatic pressures on temperature preference behavior in small aquatic animals. *Comp. Biochem. Physiol.* **68**:501–505.

Kinnunen, P.K.J., and Virtanen, J.A. 1986. A qualitative molecular model of the nerve excitability. In: *Modern Bioelectrochemistry* (F. Gutmann and H. Keyzer eds.), Plenum, New York, pp. 457–479 (Chap 17).

Koltsoff, N.K. 1927. Physicochemical fundamentals of morphology. Reprinted in 1968 In: *The Classics of Soviet Genetics*. Leningrad. pp. 85–92. (In Russian)

Laggner, P., Lohner, K., Degovics, G., Muller, K., and Schuster, A. 1987. Structure and thermodynamics of the dihexadecylphosphatidylcholine-water system. *Chem. Phys. Lipids*. **44**:31–60.

Lee, A.G. 1977. Lipid phase transitions and phase diagrams. II. Mictures involving lipids. *Biochim. Biophys. Acta* **472**:285–344.

Lohner, K., Schuster, A., Degovics, G., Muller, K., and Laggner, P. 1987. Thermal phase behaviour and structure of hydrated mixtures between dipalmitoyl- and dihexadecylphosphatidylcholine. *Chem. Phys. Lipids*. 44:61–70.

Lopez-Garcia, F., Villalain, J., Gomez-Fernandez, J.C., and Quinn, P. 1994. *Biophys. J.* **66**: 1991–2004.

Macdonald, A.G. 1984. The effects of pressure on the molecular structure and physiological functions of cell membranes. *Philos. Trans. R. Soc. Lond. B. Biol. Sci.* **304**:47–68.

Maddis, R. 1983. The evolution of sleep. In: *Sleep Mechanisms and Function*. (A. Mayes, ed.) Van Nostrand Reinhold, Cambridge, UK, pp. 57–106.

Mednikov, B.M. 2001. N.V. Timofeeff-Ressoffsky and the axiomatics of theoretical biology. *Proceedings of the Conference of Joint Institute for Nuclear Research*. Dubna, Russia, pp. 283–294. (In Russian).

Melchior, D.L., and Steim, J.M. 1976. Thermotropic transitions in biomembranes. *Annu. Rev. Biophys. Bioeng.* **5**:205–238.

Meyer, H. 1899. Zur Theorie der Alkoholnarkose. Erste Mittheilung. Welche Eigenschaft der Anästhetica bedingt ihre narkotische Wirkung? *Arch. Exp. Pathol. Pharmakol.* **425**:109–118.

Miller, K.W., Paton, W.D., Smith, R.A., and Smith, E.B. 1973. The pressure reversal of general anesthesia and the critical volume hypothesis. *Mol. Pharmacol.* **9**:131–143.

Moshkov, D.A. 1985. *Adaptation and Ultrastructure of Neuron*. Nauka, Moscow. (In Russian)

Mouritsen, O.G. 2005. *Life—As a Matter of Fat: The Emerging Science of Lipidomics*. Springer-Verlag, Berlin, 276p.

Nasonov, D.N., and Alexandrov, V.Ja. 1949. *Reaction of the Living Substance to External Actions*. USSR Acad. Sci. Publisher, Moscow, Leningrad. 252p. (In Russian)

O'Connor, C.S., Crawshaw, L.I., Bedichek, R.C., and Crabbe, J.C. 1988. The effect of ethanol on temperature selection in the goldfish, Carassius auratus. *Pharmacol. Biochem. Behav.* **29**:243–248.

Overton, C.E. 1901. *Studien über die Narkose*. Jena, Germany.

Papahadjopoulos, D., Jacobson, K., Nir, S., and Isac, T., 1973. Phase transitions in phospholipid vesicles. Fluorescence polarization permeability measurements concerning the effect of temperature and cholesterol. *Biochim. Biophys. Acta* **31**:330–348.

Pastukhov, Yu.F., Ekimova, I.V., Nozdrachev, A.D., Guselnikova, E.A., Sedunova, E.V., and Zimin, A.L. 2001. The states of sleep significantly contribute to both cooling and heating of pigeon's brain. *Dokl. Akad. Nauk* **376**:836–840. (In Russian)

Parmeggiani, P.L. 1995. Brain cooling across wake-sleep behavioral states in homeothermic species: an analysis of the underlying physiological mechanisms. *Rev. Neurosci.* **6**:353–363.

Pollack, G.H. 2001. *Cells, Gells and the Engines of Life*. Ebner & Sons, Seattle WA, USA. 320p.

Prosser, C.L. 1973. Temperature. In: *Comparative Animal Physiology.* (C.L. Prosser ed.) W.B. Saunders Company, Philadelphia-London-Toronto, (Chap 9).

Shnoll, S.E. 1979. *Physicochemical Factors of Biological Evolution.* Nauka, Moscow. English edition: Shnol, S.E. *Physicochemical Factors of Biological Evolution.* Hardwood Academic Publishers GmbH, The Netherlands, 1981 (Chap 11).

Shnoll, S.E. 2001. *Heroes, Villains, and Conformists of Russian Science.* Kron-Press, Moscow. 875p. (In Russian).

Smith, R.A., Dodson, B.A., and Miller, K.W. 1984. The interactions between pressure and anaesthetics. *Philos. Trans. R. Soc. Lond. B. Biol. Sci.* **304**:69–84.

Troshin, A.S. 1956. *Problems of Cell Permeability.* English translation: Pergamon Press, Oxford, Translated by W.F. Widdas.

Troshin, A. 1966. *Problems of Cell Permeability.* Pergamon Press, Oxford.

Trudell, J.R., Payan, D.G., Chin, J.H., and Cohen, E.N. 1975. The antagonistic effect of an inhalation anesthetic and high pressure on the phase diagram of mixed dipalmitoyl- dimyristoylphosphatidylcholine bilayers. *Proc. Natl. Acad. Sci. U. S. A.* **72**:210–213.

Ubbelohde, A.B. 1965. *Melting and Crystal Structure.* Clarendon Press, Oxford, (Chaps 11 and 14).

Ueda, I. 2001. Molecular mechanisms of anesthesia. *Keio J. Med.* **50**:20–25

Uchida, K., Yao, H., and Ema, K. 1997. Effect of chain length on the heat-capacity anomaly at the gel to liquid-crystalline phase transition in unilamellar vesicles of phosphatidylcholines. *Phys. Rev. E*, **56**:661–666.

Webb, W.B. 1983. Theories in modern sleep research. In: *Sleep Mechanisms and Function.* (A. Mayes, ed.) Van Nostrand Reinhold, Cambridge, UK, pp. 1–17.

Ephemeral Gels: The Biological Example Applied to a New Type of Polymers

J. Picard, S. Giraudier and V. Larreta-Garde

Abstract In several biological processes, a solid gel phase due to both physical and chemical bonds is produced which later dissolves. The relative influence of these two different kinds of links towards the gel properties were explored in a model system consisting of gelatin gels. Thermally reversible physical bonds, occurring with a coil to triple helix transition of gelatin chains, were obtained by cooling the protein solution. Covalent bonds were enzymatically catalyzed by a bacterial transglutaminase. Different temperature protocols were applied to gelatin solutions, with and without the enzyme, to vary the history of the gel so that different types of networks leading to various gel properties were obtained. Then, protease (thermolysin) and transglutaminase were considered as two antagonistic enzymes, one generating and the other cleaving covalent bonds. Alternate sol/gel and gel/sol transitions should occur within such a system, generating ephemeral gel phases. The various gels obtained were programmed to dissolve after a determined time, without any change in temperature or medium composition, and constitute a completely new type of material. Varying temperature, but also the ratio of the two antagonistic enzymes and the type of protease, we were able to generate a full range of ephemeral gels with controlled life times and mechanistic properties.

Keywords Gelatin gels · enzymes · dynamics · phase transition

1 Introduction

At a macroscopic scale, many biological media display gel characteristics. This soft matter state confers special properties (de Gennes, 1985) to living systems allowing dynamical behaviours. Extracellular matrix (ECM) is the best example: the solid part of the gel both constitutes a physical barrier separating organisms into tissue compartments and plays a major role in cell adhesion while the liquid phase favours the diffusion of molecules allowing intercellular communication. In

V. Larreta-Garde
Errmece, University of Cergy Pontoise, BP 222, 95302 Pontoise, France

G.H. Pollack, W.-C. Chin (eds.), *Phase Transitions in Cell Biology*,

healing, a gel is formed which serves to fill the wound. In blood coagulation, the primary clove for platelet aggregation is a soft gel. In all cases, biological natural gels are complex systems constituted of various macromolecules bound by several types of bonds. Moreover, the dense networks composing biogels in living systems are not static systems. They continuously undergo dynamic changes in composition leading to important variations at the macroscopic scale, sometimes including phase transitions (Aeschlimann and Thomazy, 2000; McCawley and Matrisian, 2001). For example, in wound healing, the first basis matrix is modified by cells which both degrade the preliminary fibrin gel and replace it by other proteins such as collagen able to organize into a different kind of supramolecular assembly. In this case, a transient gel is thus formed, undergoing a first sol/gel transition followed by other phase transitions. Extracellular matrix is also undertaking permanent remodelling, both in normal maintenance and various pathologies. In metastasis dissemination, tumor cell must cross several ECM layers that are normally impermeable to most cell types, this is achieved by locally dissolving the matrix gel. On the contrary, the links in ECM are reinforced in fibrosis giving rise to a very hard tissue (gel). Biological structures are dynamic systems not only at the molecular scale, but also at the macroscopic scale.

Whereas gels in chemistry are usually obtained through modification of either a physical parameter or by chemically inducing covalent bonds, biological gels are obtained and modified through enzyme reactions.

Enzymes usually catalyze very specific reactions often implying one single chemical group (Brenda Enzyme Database, 2007). One example is milk coagulation induced by the hydrolysis of the single $Phe_{105}-Met_{106}$ peptidic bond of κ-casein (Vanderporten and Weck, 1972). However, in some cases, these subtle modifications at the molecular scale cause important variations at the supramolecular level, sometimes generating critical phenomena leading to phase transitions. Among the 3500 characterized enzymes, a few are particularly efficient at catalysing these reactions with macroscopic effects.

These include proteases which may dissolve insoluble protein phases (Berry et al., 2000). Proteases, acting as destructive catalysts, are responsible for various physiological and pathological events (Lemaitre and D'Armiento, 2006, Deryugina and Quigley, 2006; Martisian et al., 2006) where proteolysis disorganizes the biological tissue. In contrast, proteases contribute to the generation of a new solid phase in wound healing or blood coagulation. Through zymogen activation, they generate a viscous solution or a soft gel phase which is later stabilized by transglutaminases. This second type of enzymes create covalent intermolecular protein cross-linking (Nio et al., 1986) giving rise to irreversible gel networks. Here, the two types of enzymatic reactions coexist and participate synergistically (Haroon et al., 1999; Ariens et al., 2002) to the formation of a solid phase from macromolecules in solution. However, in most biological events, these two enzymes can be considered as antagonistic ones as proteases cleave peptide bonds while transglutaminases generate isopeptide bonds. At a supra-molecular level, as far as the physical state of a proteic gel is concerned, transglutaminase contributing to the insolubilization of the protein lattice may be considered as a reverse protease. In the extracellular matrix

(ECM) remodelling (Aeschlimann and Thomazy, 2000; Van Saun and Matrisian, 2006), these two types of enzymes act conversely. Oscillating concentrations for soluble proteins and insoluble lattices were calculated in a model considering proteases and transglutaminases as antagonistic catalysts of a futile cycle in which ECM proteins and their soluble proteolysis fragments are interconverted (Larreta-Garde and Berry, 2002).

In vitro, phase transitions are due to a change in the medium composition or in the physicochemical parameters, rather than to enzymes. The aim, here, was to reproduce *in vitro* the same scheme as observed *in vivo* where the two types of reactions coexist and a solid phase is first obtained from macromolecules in solution and is later dissolved. We suggest the use of two antagonistic enzymes to catalyze sol/gel followed by gel/sol transitions in order to obtain the successive polymerization and dissolution of proteins without further addition of reactants and without any change in temperature or pressure. Here, the transglutaminase is considered as a "sol-gelase" enzyme while protease acts as a "gel-solase". This presupposes the capacity to program the global behaviour of a mixture containing at least one protein and two enzymes.

The aim of the work is to produce *in vitro* ephemeral gels mimicking the dynamics of natural transient supramolecular assemblies. To realize such purpose, it is first necessary to characterize and control the properties of various gels, then to master their hydrolysis, third to make sure that theoretically a system exhibiting a double transition may occur, finally to try to achieve it experimentally. These different stages are described in the present chapter.

2 Obtaining and Characterization of Various Gels

Among the proteins able to give rise to a gel, gelatin has the ability to form thermally reversible networks (Djabourov, 1988). Below the sol-gel transition temperature, part of the protein coils gives rise to triple helices reminiscent of the native collagen (Hulmes, 1992) and protein solution turns into gel. Physical protein gels may be stabilized by the further addition of covalent bonds due to transglutaminase reaction. Physical, Chemical, Chemical – Physical and Physical – Chemical gels were obtained by cooling the protein solution and/or by transglutaminase reaction.

First, the relative influence of physical and chemical bonds to overall gel properties of gelatin gels were characterized. Using rheology and polarimetry (Giraudier et al., 2004), the differences between physical gels and chemical gels were explored. When a gelatin solution is cooled, protein coils locally assemble into triple helices and a network is formed giving rise to a gel, the creation of physical bonds and the emergence of the resulting network were followed through the viscoelastic properties of the gel (Fig. 1 squares, Table 1). The triple helix amount increased to 15% in 150 min after the beginning of the cooling ramp (Table 1), a gel being obtained for 9% helices. Raising back the temperature leads to a gel-sol transition (Fig. 1).

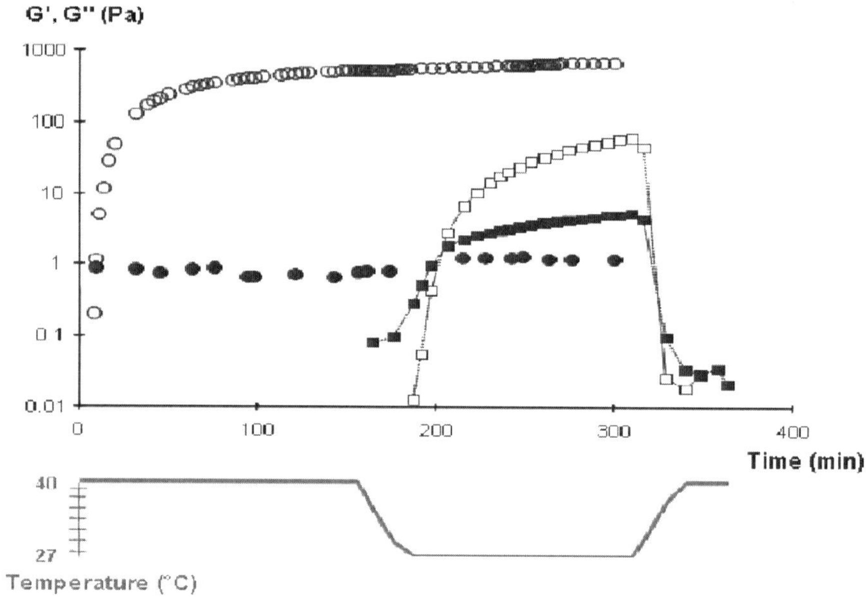

Fig. 1 Storage (*open symbols*) and loss modulus (*filled symbols*) measured at 40, 27 then 40 °C as a function of time for Physical gel (*squares*) and Chemical then Chemical – physical gel (*circles*). Gelatin concentration is 5% (W/V). Reprinted with permission from Giraudier Biomacromolecules, 5. Copyright 2004 American Chemical Society

For the chemical gel, the time dependence of the storage modulus (Fig. 1, circles) indicates a fast sol-gel transition due to the rapid formation of intra- and inter-molecular isopeptidic bonds catalyzed by transglutaminase. No helix was formed at 40 °C, then, the temperature was decreased to 27 °C and helix formation was observed in both physical and chemical gels (Table 1). For the chemical gel, the presence of a covalent network before helix formation strongly limited, but did not totally prevent triple helix propagation. This indicates that macromolecular dynamics was allowed inside the covalent gel. The obtained gel is a Chemical-Physical gel.

Table 1 Viscoelastic properties and helix % in different types of gel

Gel type	T gase, Temperature	G′ (Pa) at 150 nm	Bonds
Physical	−, 27 °C	60	Triple helix (15% after 150 min)
Chemical	+, 40 °C	527	Covalent bonds
Chemical-Physical	+, 40 → 27 °c	722	Covalent bonds + Triple helix (3%)
Physical-Chemical	+, 27 °C	907	Covalent bonds + Triple helix (9%)

Source: Reprinted with permission from Giraudier Biomacromolecules, 5. Copyright 2004 American Chemical Society

Finally, the enzymatic reaction was performed at 27 °C where coils undergo conformational transition and form triple helices so that a "Physical – chemical" gel due to both weak interactions and covalent bonds was obtained. The properties of these different gels were compared (Table 1).

To estimate gel time, t_0 was considered as the beginning of the cooling ramp. For the Chemical and Physical – chemical gels, t_0 corresponds to transglutaminase addition. Gel time was estimated when $G' = G''$.G', G'' and helix% were measured at $t_{150\ min}$.

Then, the transglutaminase concentration was varied (Fig. 2). The higher the transglutaminase concentration, the faster the sol/gel transition and the higher the elasticity of the gel. From a meticulous analysis of the curve, it was shown that despite the viscosity increase in the solution, the transglutaminase reaction is not diffusion-controlled before the sol/gel transition while the evolution of macroscopic properties is strongly reduced after the gel point. A non linear relation was obtained in the gel phase between the enzymatic reaction and enzyme concentration which is characteristic of diffusional constraints (Giraudier and Larreta-Garde, 2007).

This reveals that as the covalent bonding increases inside the gel, the diffusion of the enzyme inside the protein network becomes increasingly limited. This diffusion-reaction mechanism was quantified and taken into account for further calculations and experiments (Giraudier and Larreta-Garde, 2007).

The above results show that covalent bonding may occur simultaneously to the coil to helix transition. A comparison with the Chemical-physical gel confirms the importance of thermal history on gel characteristics. This particular gel was connected by weak interactions and covalent bonds, both having a significant influence

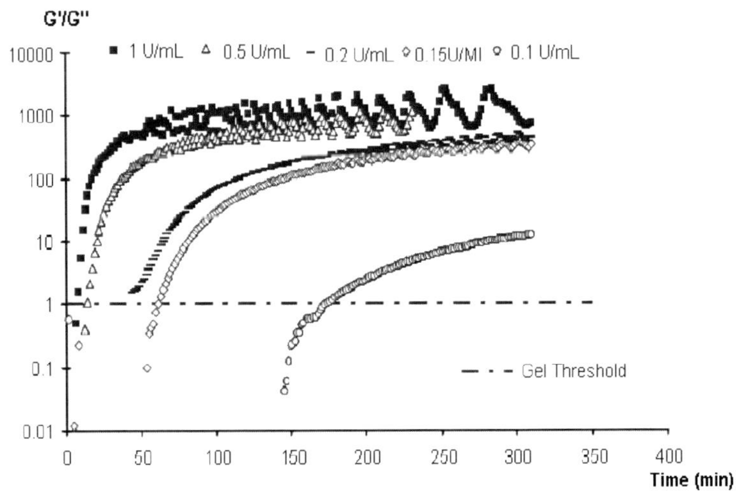

Fig. 2 G'/G'' as a function of time for different transglutaminase concentrations. (■) 1 $U.mL^{-1}$), (Δ) 0.5 $U.mL^{-1}$, (■) 0.2 $U.mL^{-1}$, (◇) 0.15 $U.mL^{-1}$, (○) 0.1 $U.mL^{-1}$. A gel is obtained for $G'/G'' > 1$. Gelatin concentration is 5% (W/V). Reprinted with permission from Giraudier Macromolecular symposia 256. Copyright 2007 Wiley-VCH

on network viscoelasticity. The dynamics of helix directly depends on covalent association of gelatin chains by transglutaminase. From these observations, the following scheme may be proposed to describe the different supramolecular organizations of the various gels.

It has been shown that the overall properties as well as the dynamics inside the gels are dependent upon the order of formation and on the relative amount of triple helices and covalent bonds (Giraudier et al., 2004). The next stage was to consider this internal organization of the gel towards proteolysis.

3 Gel Hydrolysis

The differences in the gel networks have been correlated to the variations in hydrolysis rates by thermolysin (Fig. 3). A gel is realized in the bottom of a cuvette and a solution containing a protease covers the gel. Optical density at 280 nm is followed (Giraudier et al., 2004). To estimate the temperature influence on thermolysin activity, Vi was measured at both at 27 and 40 °C on a reference substrate (Berry et al., 2000), a $Vi_{40^\circ C}/Vi_{27^\circ C}$ ratio determined (1.66) and applied to Chemical gel hydrolysis (performed at 40 °C).

The enzymatic hydrolysis of the various gels give additional information about their structure and organization. Hydrolysis of the Physical gel occurs faster than that of the covalently linked gels, confirming that transglutaminase cross-linked products are more resistant to proteolytic degradation. Additional information about the various networks and their behaviour towards proteases was obtained from the measurement of helix release and degradation by polarimetry (Giraudier S. et al., 2004).

From these preliminary results, it has been confirmed that the history of the gel is of crucial importance for its macroscopic properties. Moreover, a network with

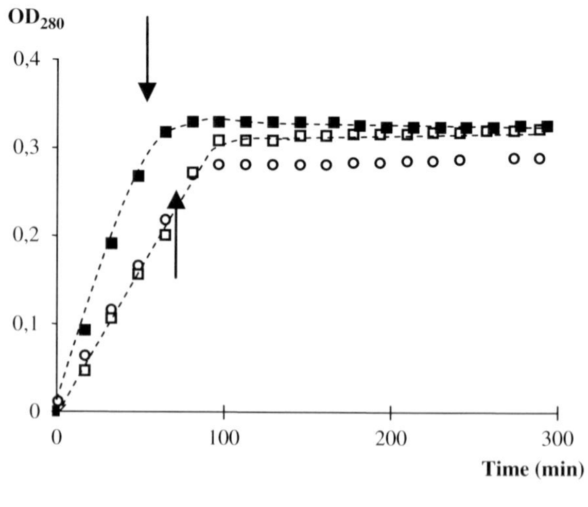

Fig. 3 Release of degradation products in the liquid phase as a function of time. Hydrolysis was carried out by 5 μM thermolysin on Physical (*filled squares*), Physical – chemical gel (*open squares*) and Chemical (*open circles*, standardized curve) gelatin gels. Arrows indicate total gel degradation. Reprinted with permission from Giraudier Biomacromolecules, 5. Copyright 2004 American Chemical Society

both helices and covalent bonds could be protected from the enzymatic hydrolysis which is of biological and industrial relevance.

4 Ephemeral Gels

After the hydrolysis of the gels with a thermolysin added outside a previously formed gel, the protease was added inside the gelatin solution before the formation of a physical gel (Fig. 4).

A gel is rapidly obtained after the gelatin solution is cooled, then under the protease action the gel is dissolved. Polarimetry measurements show that helices are not destroyed by thermolysin reaction. A second cooling of the hydrolyzed solution at 20 °C allows the gel to form again, this gel/sol transition is concomitant with the appearance of new helices. This experiment shows that experimentally a gel can undergo alternate gelation transitions when modifying the physicochemical conditions and adding a protease.

The aim was to avoid the modification of physico-chemical parameters or medium composition to generate the gel/sol transition. The next stage of this study is to program the macroscopic properties of the gel, i.e. the phase transitions from the formation of the network by the use of convenient enzymes. The concomitant presence of two antagonistic enzymes, i.e. transglutaminase and thermolysin is considered. In a first stage, a theoretical model was elaborated to demonstrate the possible occurrence of transient gel states.

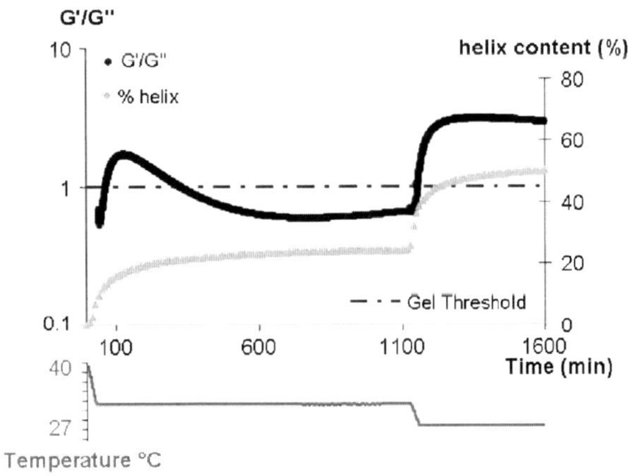

Fig. 4 Viscoelastic behaviour and helix content of a physical gel containing a protease as a function of time. The temperature protocol is as follows: 40 °C → 27 °C → 20 °C. A gel is obtained for $G'/G'' > 1$. Gelatin concentration is 5% (W/V). Reprinted with permission from Giraudier Macromolecular symposia 256. Copyright 2007 Wiley-VCH

5 Kinetic Model

Two enzyme reactions are considered. The first one is catalyzed by a transglutaminase (T) which binds soluble protein molecules (s) into a protein chain network (g). The second reaction is catalyzed by a protease (P) and dissolves the bound proteins. In our model, protease and transglutaminase become the two antagonistic catalysts of a futile cycle between a protein network and its proteolysis fragments. At a macroscopic scale, this futile cycle also represents a double phase transition: from sol to gel under the transglutaminase action and from gel to sol under the protease action. In addition, the protease hydrolyzes soluble proteins into fragments (f) which are too small to incorporate into the protein network, resulting in fragment efflux from the cycle. A schematic representation of our model is shown in Fig. 5-A. Below the sol-gel transition temperature, part of the protein coils gives rise to triple helices reminiscent of the native collagen and protein solution spontaneously turns into gel. When the enzymatic reaction was performed at 27 °C where coils undergo conformational transition and form triple helices so that a "Physical – chemical" gel due to both weak interactions and covalent bonds was obtained, the triple helix formation (non enzymatic reaction) was added to the model (dashed line in Fig. 5-B).

Both enzymes are assumed to display Michaelis-Menten kinetics with k_p and k_T as the catalytic constants and K_P and K_T the Michaelis constants for protease and transglutaminase, respectively. Protease-catalyzed degradation of the fragments is also taken into account. The affinity of the protease is supposed to be identical for soluble and insoluble proteins (*i.e.* K'_p and K_P). P and T are the concentrations of protease and transglutaminase respectively, which remain constant. Diffusional constraints were quantified and integrated in the model, the details of which have been described elsewhere (Giraudier and Larreta-Garde, 2007). For V_H calculations, the fact that three gelatin chains are needed to form a triple helix was taken into account.

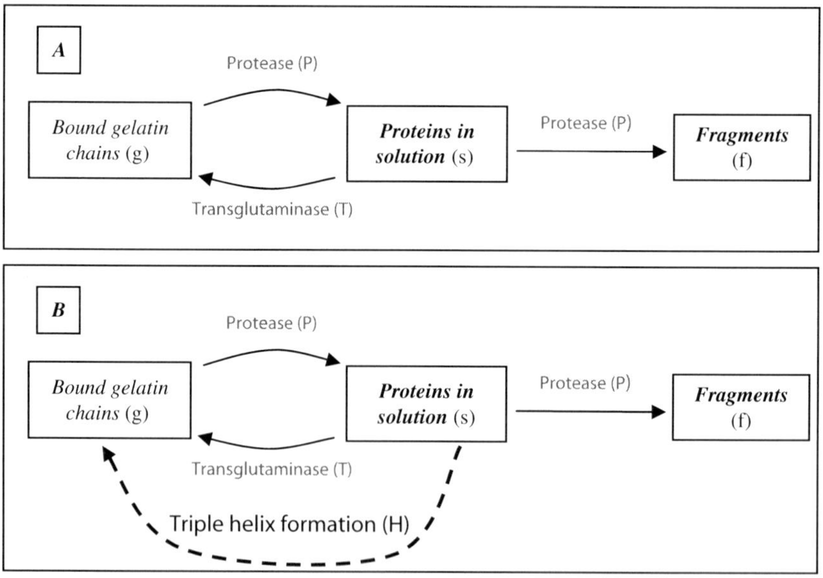

Fig. 5 Futile cycle between a protein network and its proteolysis fragments using protease and transglutaminase as antagonistic catalysts. **A**: reactions at 40 °C, **B**: reactions at 27 °C. Reprinted with permission from Giraudier Biophys. J. 93 (2). Copyright 2007 Biophysical Sciety

Designating g, s and f, as the time-dependent concentrations of bound gelatin chains, soluble proteins, and their corresponding fragments, respectively, T, the transglutaminase concentration, and P the protease concentration, a set of three ordinary differential equations is obtained*:

$$\frac{dg}{dt} = -\frac{k_P P}{K_P + g} g + \frac{k_T T^{\beta}}{K_T + s} \mathrm{s}\,(+V_H) \quad (1)$$

$$\frac{ds}{dt} = \frac{k_P P}{K_P + g} g - \frac{k_T T^{\beta}}{K_T + s} \mathrm{s} - \frac{k'_P P}{K_P + s} s(-V_H) \quad (2)$$

$$\frac{df}{dt} = \frac{k'_P P}{K_P + s} s \quad (3)$$

$$g_t + s_t + f_t = s_0 \quad (4)$$

The evolution versus time of the bound gelatin chains (g) was expressed as a percentage of the total protein concentration. An example is shown in Fig. 6 for a ratio of transglutaminase/protease activity of 10 at 27 °C where the effect of triple helices added to covalent bonding.

These theoretical results shows that the rate of bound protein chains increases over the threshold needed for a sol/gel transition and after a while, decreases spontaneously so that a solution should be obtained again. The kinetics as well as the overall properties of the gels depend on the relative enzyme activities and on operational conditions.

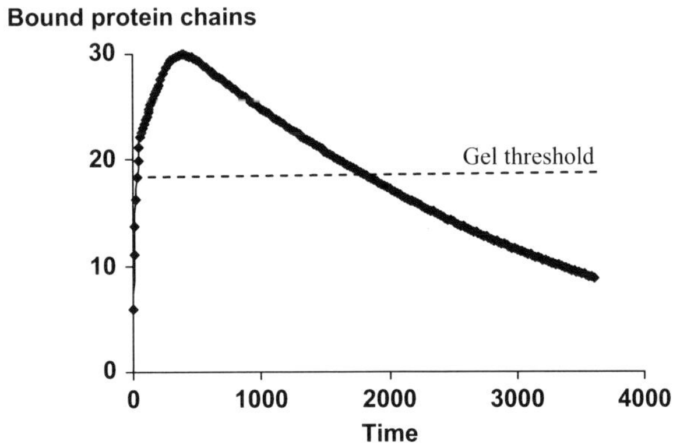

Fig. 6 Theoritical evolution of gelatin bound chains (%) as a function of time (adimensional) at 27 °C with $k'_P = 0.05\ k_P$ and $k_T \cdot T/k_P$ P = 10. Reprinted with permission from Giraudier Biophys. J. 93 (2). Copyright 2007 Biophysical Society

* The equation 1, 2, 3, 4 Reprinted with permission from Giraudier Biophys. J. 93 (2). Copyright 2007 Biophysical Society

For example, at 27 °C, gelation should occur faster and more chains be bound than at higher temperatures so that more stable gels would be observed; the double transition also needs less transglutaminase to occur in this case. From this model, we determined enzyme activity ratios where the double transition could occur.

Gels were then realized experimentally in the presence of the two antagonistic enzymes, i.e. transglutaminase and thermolysin using the conditions defined through the theoretical model. In the given example (Fig. 7), the two enzyme reactions were carried out at 40 °C, a temperature where no physical gelatin gel can be obtained, so that only transglutaminase-catalyzed covalent bonds are responsible for the gelation process. The properties of the gel were followed by rheology, as previously described (Giraudier et al., 2004). Under these conditions, a gel is obtained when the storage modulus G' becomes higher than the loss modulus G″, that is to say when $G'/G'' > 1$. At a constant gelatin concentration (50 g. L^{-1}), various transglutaminase to protease ratios (T/P) were tested and also the enzyme activity was varied for a constant T/P ratio. It is important to mention that during the whole experiment, the temperature is kept constant.

The analysis of the results on Fig. 7 indicates that the gelation process is an accurate phenomenon under these conditions. With a T/P ratio of 75 no gel was obtained (the experiment was carried out over 80 h), while increasing this value to 80 allowed gelling to proceed. This first phase transition occurs in 19 min, which is slower than for a gel without protease (8.5 min). As predicted by a mathematical model (Giraudier and Larreta-Garde, 2007), the gel is evolving: its viscoelasticity increases to reach a maximal value ($G'/G'' = 26$) in 165 min, then dissolution begins, being achieved in 760 min. The process is irreversible, no spontaneous gelling was observed beyond this point. This result is consistent with the hypothesis of small fragments development due to the protease reaction on soluble gelatin chains.

The overall behaviour of the preparation is also strictly dependent on the enzyme activity. A slight decrease in the enzyme concentrations (0.4 unit instead of 1 unit transglutaminase, T/P ratio constant) results in an increase in the gelling time

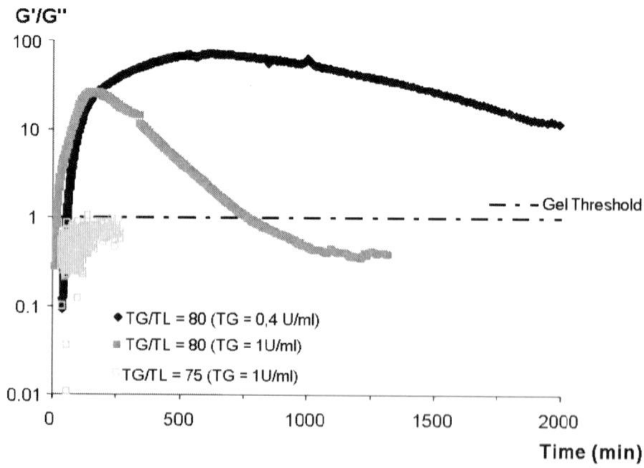

Fig. 7 Evolution of the viscoelastic properties of ephemeral gels as a function of time (min) at 40 °C for T/P = 80 with T = 0.4 $U.mL^{-1}$(♦) or T = 1 $U.mL^{-1}$ (■) and for T/P = 75 with T = 1 $U.mL^{-1}$ (●). A gel phase is obtained when $G'/G'' > 1$. Gelatin concentration is 5% (W/V). Reprinted with permission from Giraudier Biophys. J. 93 (2). Copyright 2007 Biophysical Society

(56 min instead of 19 min), higher visco-elastic properties ($G'/G'' = 71$ after 10 h), and a longer gel phase (dissolution needs 80 h).

A physical-chemical ephemeral gel was realized at 27 °C. At this temperature, the enzymatic reaction is concomitant with the triple helix formation. The obtained gel is due to both weak interactions and covalent bonding. Varying temperature, completely different properties were observed (Fig. 8).

At a temperature where both triple helices and covalent bonds contribute to the network, the gel resistance is increased. These results are consistent with the ones previously observed when adding protease outside the gel (Giraudier et al., 2004). One may remark that helices are not degraded during the gel/sol transition. This is due to the particular choice of the protease, thermolysin which preferentially recognizes hydrophobic residues with large lateral chains (Matsubara, 1996), as they are rare in helices, thermolysin should preferentially act on the random coil part of the gel chains.

Next, an ephemeral gel was realized at 40 °C based on a chemical gel, and then, after the gel/sol transition the obtained solution was cooled to 27 °C to induce the formation of triple helices (Fig. 9).

We can observe that, in this case, despite the occurring of an amount of triple helices sufficient to induce a sol/gel transition (9%), no physical gel was obtained after the dissolution of the chemical gel. This particular effect is due to the high rate

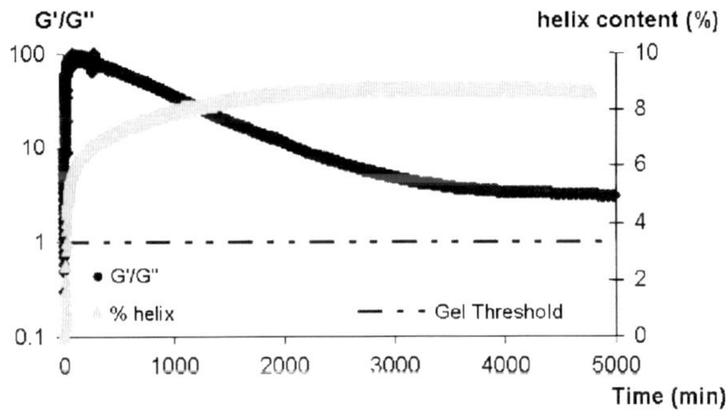

Fig. 8 Evolution as a function of time of the viscoelastic properties and triple helix content of a physical-chemical ephemeral gel at 27 °C for T/P = 80. Gelatin concentration is 5% (W/V). Reprinted with permission from Giraudier Biophys. J. 93 (2). Copyright 2007 Biophysical Society

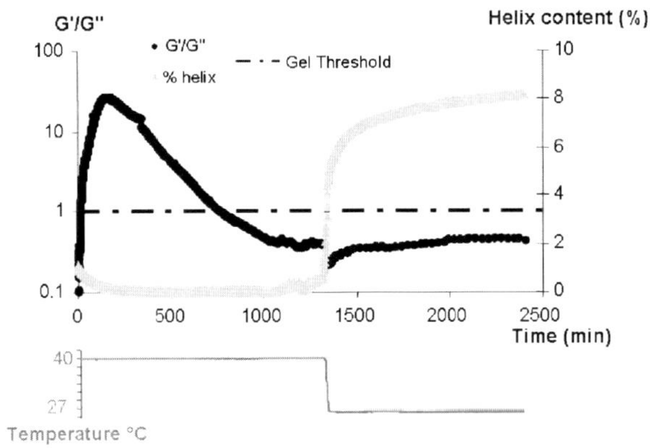

Fig. 9 Evolution as a function of time of the viscoelastic properties and triple helix content of a chemical ephemeral gel thus cooled at 27 °C. T/P = 80. The thermal protocol is 40 °C → 27 °C. Reprinted with permission from Giraudier Macromolecular symposia 256. Copyright 2007 Wiley-VCH

of hydrolysis realized by thermolysin. The medium contains too many small protein chains and peptides to allow the obtaining of a correct network although the molecular mobility of the protein chains allows the triple helix association. This result brings new insights on the relation between triple helices and gelation. Contrarily to what was observed with native gelatin (Joly-Duhamel et al., 2002), there is no direct relation in the present case between triple helix content and G′. These results illustrate that varying the temperature, we were able to generate ephemeral gels with various life times and mechanistic properties. A full range of controlled properties were obtained by modifying the protein nature and concentration. We thus tested the role of protease specificity.

The second protease used was a collagenase which is specific of the Gly-X-Y sequence responsible for the triple helix formation.

When different gels were hydrolyzed with collagenase, the same hydrolysis rate was observed whatever the type of bonds inside the gel (Fig. 10). This shows that for this enzyme, contrarily to what was observed with thermolysin, covalent bonds do not protect the network against the protease. Moreover, the formation of triple helices is not important for collagenase as the reaction rate is identical whether helices are formed (physical gel) or not (chemical gel). The main difference between the two enzymes is that collagenase mainly reacts inside the helix sequence while thermolysin acts outside it, where the covalent bonds are localized (Giraudier et al., 2007). So, protease specificity is of crucial importance for the hydrolysis of a protein gel network. It was then studied on ephemeral gels (Figs. 11 and 12).

Different properties were obtained with the two proteases. In the presence of collagenase, less triple helices are formed which are hydrolyzed by the protease. The gel obtained with collagenase is much more elastic than the one obtained with thermolysin and its profile is close to that of a chemical one. In this last case, collagenase cleaves helices, but the hydrolyzed chains are still locally associated and bound by covalent bonds network (see Scheme 1), a further hydrolysis is thus needed to solubilize the gel. With thermolysin, the hydrolysis does not affect the helix structures, but mainly the random chains bound by covalent bonds, so that one

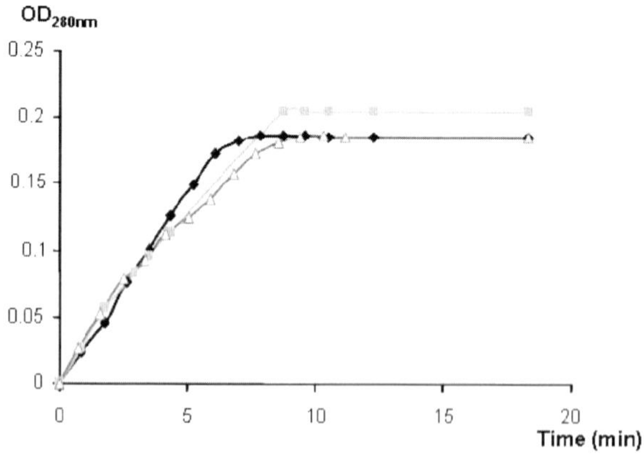

Fig. 10 Hydrolysis by collagenase of a physical gel (♦), a chemical gel (Δ) or a physical-chemical gel (▒) as a function of time. Gelatin concentration is 7% (W/V). Reprinted with permission from Giraudier Macromolecular symposia 256. Copyright 2007 Wiley-VCH

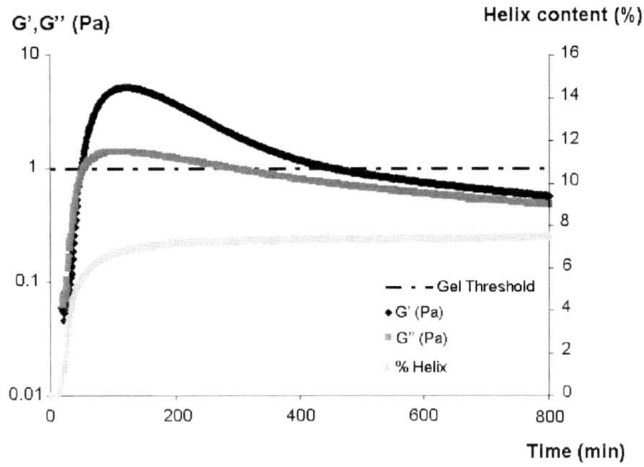

Fig. 11 Ephemeral physical–chemical gel behaviour with thermolysin as a function of time. G′ (♦), G″ (■) and helix % (▲) were followed. Reprinted with permission from Giraudier Macromolecular symposia 256. Copyright 2007 Wiley-VCH

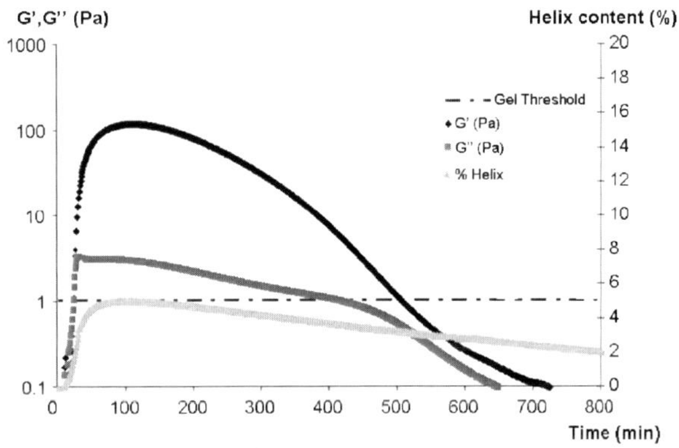

Fig. 12 Ephemeral physical–chemical gel behaviour with collagenase as a function of time. G′ (♦), G″ (■) and helix% (▲) were followed. The reaction is carried out at 27 °C. Gelatin concentration is 7% (W/V). Reprinted with permission from Giraudier Macromolecular symposia 256. Copyright 2007 Wiley-VCH

single cleavage may strongly destabilize the network. The differences in molecular recognition of the protein by the enzymes are of fundamental importance for the macroscopic properties of the gel and for its global behaviour.

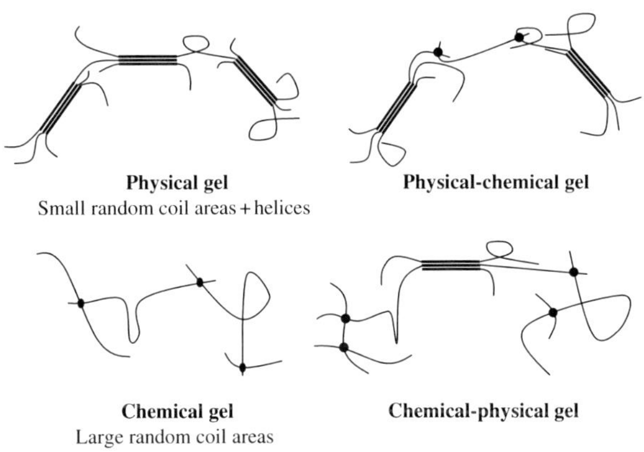

Scheme 1 Proposed organization of the various gels. Reprinted with permission from Giraudier Biomacromolecules, 5. Copyright 2004 American Chemical Society

6 Conclusion

An increasing amount of work has been dedicated in the past years to phase transitions in biological media. Some tissue or organs behave like gels; phase transitions are then related to main physiological or severe pathological behaviours such as wound healing or tumour invasiveness. Enzymes are usually responsible for the related sol/gel and/or gel/sol transitions. However, the consideration of enzymes to the macroscopic variations of cell environment has mainly been restricted to the catabolic action of proteases. Our main objective in this chapter consisted in the study of alternate phase transitions due to antagonistic enzyme activities as observed in matrix remodelling. We added transglutaminases to a protein solution as these wide-spread enzymes can turn soluble proteins into insoluble lattices; their influence on the mechanical properties of biological gels appears to be the reverse of the protease action. Proteases and transglutaminases were treated as reverse catalysts of a futile cycle in which protein gels and their soluble proteolysis fragments are interconverted. Using gelatin, we first focused on the kinetics of the two antagonistic enzymes as far as sol/gel and gel/sol transitions are concerned.

Then we suggested the possibility to obtain ephemeral gels presenting alternate phase transitions. First, a mathematical model based on a transglutaminase/ protease cycle interconverting soluble proteins and an insoluble network was elaborated. This model allows the prediction of protein bonding/dissociation and of the resulting phase transitions (Giraudier and Larreta-Garde, 2007). Then based on the model, we were able to obtain gelatin ephemeral gels experimentally, examples of which are described in the present paper. We thus used various proteases and obtained ephemeral gels with a large amount of properties.

We have shown here that two antagonistic enzyme activities, inside a mixed network, may generate a dynamic modification of protein network which induces a sol/gel transition followed by a gel/sol transition. Using both proteases and transglutaminases with a gelatin solution, we obtained a completely new type of material which we term "Enzgels". This is the first report in protein polymer literature where no modification in temperature or medium composition is required to dissolve a gel *in vitro*. The "Enzgels" are dynamic protein solutions able to spontaneously cycle from a sol/gel to a gel/sol transition. The interest of this type of material resides in their ability to be completely pre-programmed. The transition is depending on the conformation of proteins constituting the network, but the type of bonds forming the network itself is even more important.

By creating "Enzgels" we made a strong rupture in biomaterial field. The mastering of a double kinetic upon time (polymerisation and dissolution) opens a wide range of applications. "Enzgels" can provide a new proteic vehicle to deliver drug with fine time control. The double kinetic leads to new purposes in pharmaceutical industries, in dermal or surgery wound care market.

Moreover, these studies participate to the knowledge of enzyme behaviour in non conventional media.

Finally, the main achievement of this work was to realize *in vitro* alternate phase transitions as they sometimes occur in nature. Starting from the observation of biological examples, the knowledge here acquired may contribute to the understanding of complex biological processes such as the matrix remodelling involved in fibrosis or cancer metastasis.

References

Aeschlimann D. and Thomazy V. 2000. *Connective Tissue Res.* 41, 1–27.
Ariens R.A., Lai T.S., Weisel J.W., Greenberg C.S. and Grant. P.J. 2002. *Blood* 100, 743–754.
Berry H., Pelta J., Lairez D. and Larreta-Garde V. 2000. *Biochim. Biophys. Acta* 1524,110–117.
Brenda Enzyme Database 2007. http://www.brenda.uni-koeln.de/
De Gennes P.G. 1985. *Scaling concepts in polymer physics*, Cornell University Press, Ithaca, NY.
Deryugina E.I. and Quigley. J.P. 2006. *Cancer Metastasis Rev.* 25, 9–34.
Djabourov M. 1988. *Contemp. Phys.* 29, 273–297.
Giraudier S., Hellio D., Djabourov M., Larreta-Garde V. 2004. *Biomacromolecules* 5, 1662–1666.
Giraudier S. and Larreta-Garde V. 2007. *Biophys. J.* 93(2), 629–636.
Giraudier S., Picard J. and Larreta-Garde V. 2007. Macromolecular Symposia Polymer Networks 2006. 256(1), 158–166.
Haroon Z.A., Hettasch J.M., Lai T.S. and Greenberg. C.S. 1999. *FASEB J.* 13, 1787–1795.
Hulmes D.J.S. 1992. *Essays Biochem.* 27, 49–67.
Joly-Duhamel C., Hellio D., Ajdari A. and M. Djabourov 2002. *Langmuir* 18, 7158–7166.
Joly-Duhamel C., Hellio D. and M. Djabourov 2002a. *Langmuir* 18, 7208–7217.
Matrisian L.M., Sledge G.W. and Mohla S. 2006b. *Cancer Res.* 63, 6105–6109.
Larreta-Garde V. and Berry H. 2002. *J. Theor. Biol.*217, 105–124.
Lemaitre V. and D'Armiento J. 2006. *Res. C Embryo Today* 78, 1–10.
Matsubara H. 1966. *Biochem. Biophys. Res. Commun.* 24, 427–430.
McCawley L.J. and Matrisian L.M. 2001. *Curr. Opin. Cell Biol.* 13, 534–540.
Nio N., Motoki M. and Takinami K. 1986. *Agric. Biol. Chem.* 50, 851–855.
Vanderporten C. and Weck, M. 1972. *Milk Dairy* 26, 47–59.
Van Saun M.N. and Matrisian L.M. 2006. *Res C Embryo Today* 78, 69–79.

The Cytoskeleton of the Living Cell as an Out-of-Equilibrium System

Guillaume Lenormand, Adriano M. Alencar, Xavier Trepat, En-hua Zhou, Ben Fabry, James P. Butler and Jeffrey J. Fredberg

Abstract In a remarkably broad range of eukaryotic cell types, the cytoskeleton exhibits physical properties and remodelling dynamics strongly reminiscent of soft inert condensed systems, although with important differences as well. This unexpected intersection of condensed matter physics and cytoskeletal biology suggests that trapping, intermittency, and approach to kinetic arrest represent central mesoscale features linking underlying molecular events to integrative cellular functions such as crawling, contraction and remodelling.

Keywords Magnetic twisting cytometry · soft glassy rheology · jamming · spontaneous bead motion · actin dynamics · airway smooth muscle

Acronyms and Abbreviations

AFM	Atomic Force Microscopy
ATP	Adenosine Triphosphate
CSK	Cytoskeleton
DBcAMP	N6,2-O-dibutyryladenosine3,5-cyclic monophosphate
ECM	Extracellular Matrix
EM	Electron Micrograph
FDT	Fluctuation-Dissipation Theorem
GSER	Generalized Stokes-Einstein Relationship
HASM	Human Airway Smooth Muscle
LTM	Laser Tracking Microrheology
MA	Micropipette Aspiration
MSD	Mean Square Displacement
OMTC	Optical Magnetic Twisting Cytometry

G. Lenormand
Molecular and Integrative Physiological Sciences, Department of Environmental Health, Harvard School of Public Health, Boston, MA, USA
e-mail: glenorma@hsph.harvard.edu

G.H. Pollack, W.-C. Chin (eds.), *Phase Transitions in Cell Biology*,

SGM	Soft Glassy Materials
SGR	Soft Glassy Rheology
TPM	Two Point Microrheology
USR	Uniaxial Stretching Rheometer

1 Introduction

Mechanical properties of the adherent eukaryotic cell are largely determined by its cytoskeleton (CSK), a complex biopolymer network consisting of filamentous actin, microtubules, and intermediate filaments, all of which are associated with crosslinkers, motor proteins, and regulatory proteins. The living cell can modify its mechanical properties and numerous other biological functions in response to the stiffness of the substrate on which it adheres, as well as in response to external forces (Discher et al., 2005; Engler et al., 2006; Ingber, 2006; Janmey and Weitz, 2004). An integrated understanding of cell mechanics becomes therefore an essential first step for elucidating many fundamental aspects of cell behaviours.

In this chapter, we review compelling evidence showing that the dynamics of the adherent living cell shows surprising and unanticipated similarities with that of inert soft glassy materials (SGMs). The class of SGMs includes foams, pastes, colloids, emulsions, and slurries; though very different in structure, the dynamics and mechanical behaviour of each of these substances are surprisingly alike. All are very soft (elastic moduli in the range of $10^2 - 10^4$ Pa), and both the storage modulus, G′, and the loss modulus, G″, increase with weak power law dependencies on frequency. Moreover, the loss tangent or hysteresivity, $\eta = G''/G'$, is frequency insensitive and on the order of 0.1 (Sollich, 1998). Importantly, the near constancy of η in an SGM implies that the mechanism of mechanical energy dissipation is not coupled to a viscous stress, but rather is coupled to the elastic stress; in a manner that is remarkably insensitive to the rate of cyclic deformation, energy dissipated per cycle becomes nearly a fixed fraction of the elastic energy stored (Fredberg and Stamenovic, 1989). Furthermore, such glassy systems are characterized by unusual out-of-equilibrium slow dynamics, aging, rejuvenation, and intermittent rearrangements of internal structures, as described below.

Soft glassy dynamics can be studied most easily in a crowded colloidal suspension (Weeks et al., 2000). With sufficient crowding, particles can become trapped by their surrounding neighbours. Because these neighbours act as a cage, particle motions are localized most of the time, although occasionally a particle escapes its trap and thus causes the local configuration to rearrange (remodel). With time, however, the system evolves into configurations that are more and more stable, but this evolution occurs more slowly than any exponential process. Slowly evolving dynamics of this kind is called aging (Struick, 1978). In such systems, a sufficiently large oscillatory shear can provide an additional energy source that can overcome energy barriers and drive structural rearrangements. As a consequence, material properties return to their earlier values. In this way aging is reversed, and the system is said to be ‘rejuvenated’ (Struick, 1978).

This chapter begins with a summary of the different experimental techniques that are now used to probe CSK dynamics, with an emphasis on the strengths and weaknesses of each technique. We then go on to describe phenomenological laws that would appear to be nearly universal in the sense that they seem to capture the behaviour of a remarkably wide range of systems with very few parameters. In particular, we review evidence showing that the CSK of the living cell is soft and prestressed, and that it exhibits scale-free dynamics, physical aging and rejuvenation, and anomalous remodelling properties. We then contrast these results with the theory of soft glassy rheology (SGR). We end this chapter by highlighting a few intriguing questions.

2 A Few Tools to Measure CSK Dynamics

Since Francis Crick and Arthur Hughes' article on the physical properties of cytoplasm (Crick and Hughes,1950), various innovative techniques have been developed to probe mechanical properties of the living cell. Microrheology, i.e., measurement of mechanical properties on a microscopic length scale, can be obtained using two different approaches. Active microrheology applies an external stress (strain) on a probe and measures the resulting strain (stress). In contrast passive microrheology uses spontaneous thermally-driven motions of a probe to infer mechanical properties of the surrounding medium. We now briefly review some of the most commonly used methods and highlight their relative strengths and weaknesses.

2.1 Active Microrheology

2.1.1 Optical Tweezers, Magnetic Tweezers, and Optical Magnetic Twisting Cytometry (OMTC)

Each of these techniques can be thought of as a micro-rheometer in which the cell is deformed between a plate at the cell base (i.e. the cell culture dish on which the cell adheres) and a microbead of a few microns in diameter, usually partially embedded onto the cell surface (Fig. 1a). A mechanical load is then applied to the bead using optical tweezers, magnetic tweezers, or magnetic twisters (Balland et al., 2005; Fabry et al., 2001a, 2003; Hoffman et al., 2006; Matthews et al., 2004; Overby et al., 2005; Van Citters et al., 2006; Wang et al., 1993). As the bead moves, it deforms the cell to which it is bound, and the cell resists this bead motion by developing internal stresses that depend on the cell's mechanical properties. Thus cell mechanical properties can be inferred from the relationship between the imposed mechanical stress (or strain) and the measurement of the resulting strain (or stress).

When exerting a sinusoidal stress at the frequency f, one can measure the complex modulus, $G^*(f)$ (Findley et al., 1989). At a given frequency, the complex modulus is defined as the ratio of the stress to the resulting strain (Balland et al., 2005; Fabry et al., 2001a, 2003). The real part of G^*, G', is the storage modulus and

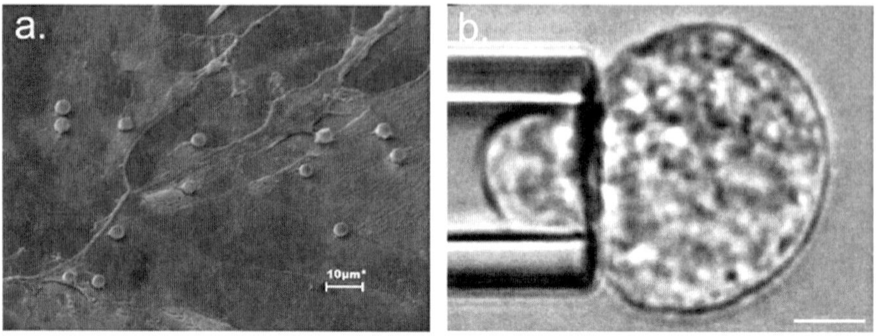

Fig. 1 Examples of active techniques. (**a**) OMTC: Electron micrograph (EM) of magnetic microbeads (5 μm in diameter) attached to the apical surface of human airway smooth muscle (HASM) cells. Scale bar is 10 μm. (**b**) Micropipette aspiration of an NIH 3T3 fibroblast. Scale bar is 5 μm

corresponds to the stiffness; the imaginary part, G'', is the loss modulus and corresponds to the friction. When exerting a constant stress, one can measure the resulting creep function, J(t) (Findley et al., 1989). The creep function is defined as the ratio of the temporally evolving strain to the constant applied stress (Bausch et al., 1998; Desprat et al., 2005; Lenormand et al., 2004). A well-defined linear range of cell deformations has been established (Fabry et al., 2001a, 2003; Lenormand et al., 2007a), and within that linear range $G^*(f)$ and J(t) contain equivalent information and are directly obtained through Laplace transformation (Lenormand et al., 2004).

Limitations and artefacts associated with all active bead-based techniques have been reported in detail elsewhere (Bursac et al., 2007; Fabry et al., 2003; Lenormand et al., 2007a; Puig-de-Morales et al., 2004). Among these limitations are that the geometry of the bead–cell interaction is not controlled, and that the interaction of the microbead with the cell induces local remodelling events that alter the structure that is being probed. It is important to recognize that while the bead has caused the cell to remodel locally, these remodeled structures are no different in kind from those that form at anchorage sites to the extracellular matrix (ECM) upon which the cell is adherent; indeed, from the point of view of the adherent cell, the bead is merely another piece of ECM upon which adhesion sites might be generated (Puig-de-Morales et al., 2004; Wang et al., 1993). As in atomic force microscopy and micropipette aspiration, described below, another limitation is that in order to obtain an elastic modulus with units of Pascals a model-dependent factor must be invoked. This factor depends on geometry, including cell height and the degree of bead embedding, and also depends upon assumed constitutive laws describing internal cytoskeletal deformations (Dimitriadis et al., 2002; Karcher et al., 2003; Mijailovich et al., 2002; Ohayon et al., 2004).

Principal strengths include the fact that the bead can apply a mechanical load to the cell body via specific receptor-ligand systems selectable by the bead coating (Puig-de-Morales et al., 2004), and that the applied stresses can span or exceed the physiologic range of stress (from below 1 Pa to over 1000 Pa). Similarly, bead displacements as small as 5 nm can be resolved quite readily (Fabry et al., 2003), and both forced and spontaneous bead displacements can be measured in the very same bead (Bursac et al., 2005). Other major strengths are the ability to perform measurements over a wide range of frequencies – from 10^{-2} to 10^{3} Hz using OMTC (Fabry et al., 2001a, 2003) – and, as described below, the capacity for massively parallel data acquisition. This latter factor becomes particularly important because of innate variability; mechanical properties of cells follow a log-normal distribution with a geometric standard deviation of about 3 (Desprat et al., 2005; Fabry et al., 2003; Hoffman et al., 2006).

2.1.2 Atomic Force Microscopy (AFM)

Atomic Force Microscopy (AFM) uses a sharp tip at the end of a flexible cantilever to indent the sample locally. The position of the tip is controlled with a piezoactuator and the force applied to the sample is obtained from the bending of the cantilever. AFM has been successfully used to probe cell mechanics over three frequency decades (0.1–100 Hz) and at different loading forces (0.1–09 nN) (Alcaraz et al., 2003; Roca-Cusachs et al., 2006; Smith et al., 2005). The sharp geometry of the tip and the high accuracy of its positioning make it possible to measure the mechanical properties of very small structures, such as dendritic spines in living neurons (Smith et al., 2007).

The major strength of AFM lies in its versatility. Cantilevers are commercially available in a wide variety of dimensions, spring constants, and tip shapes. They do not require coating to apply compressive stresses directly to the cell surface, but they can also be functionalized to measure adhesion forces. Given that the shape of the AFM indenter is well defined, the use of a suitable contact elastic model enables the determination of the cell indentation and the tip-cell contact geometry (Rico et al., 2005). Accurate control of the tip position in the plane parallel to the sample allows high-resolution imaging and rapid force mapping of the sample surface (A-Hassan et al., 1998). A major limitation of the application of AFM to cell mechanics is that the viscous drag acting on the cantilever needs to be corrected at high frequencies to obtain a reliable measurement of the cell complex modulus (Alcaraz et al., 2002). In addition, the time required to measure the rheology of a sufficiently large cell population is substantially larger than that of other techniques such as OMTC, or one-point and two-point microrheology. This is a non-trivial limitation given the large variability of the cell mechanical properties (see Section 2.3.2).

2.1.3 Uniaxial Stretching Rheometer (USR)

In a Uniaxial Stretching Rheometer (USR), a single living cell is stretched between two glass microplates, one rigid and one flexible. Displacement of the rigid microplate is achieved using a closed-loop piezoelectric translator, and the deflection of the flexible microplate allows the measurement of the applied force. USR is usually used to measure the creep function, i.e. the deformation under constant stress, or the relaxation function, i.e. the stress relaxation under constant strain (Desprat et al., 2005; Thoumine and Ott, 1997).

Principal strengths of this technique are that it probes the entire cell body, it applies a load in the physiological range, the geometry of the deformation is measurable and well defined, and very large deformations can be applied. Interestingly this technique is among the few to measure stress relaxation. The major weaknesses are that measurements are performed one single cell at a time, and that regional differences in cell properties cannot be probed.

2.1.4 Micropipette Aspiration (MA)

Micropipette Aspiration (MA) is a technique based on the principle of hydrostatic pressure transmission. The system generally comprises a glass micropipette that directly aspirates the living cell, and a hydraulic system that controls the suction pressure in the micropipette. The MA technique has been widely used to measure the force-deformation relationship of living cells in suspension, such as neutrophils (Schmid-Schonbein et al., 1981), chondrocytes (Jones et al., 1999), endothelial cells (Theret et al., 1988), fibroblasts (Thoumine et al., 1999), red blood cells (Nash and Gratzer, 1993), and even the isolated cell nucleus (Dahl et al., 2005; Rowat et al., 2006).

Viscoelastic properties of a living cell are generally measured using a creep test, where the projection length, $L_p(t)$, of the cell into the micropipette is recorded as a function of time, t, in response to a step in pressure (Fig. 1b). One can then calculate the creep function of the cell according the suction pressure, the micropipette diameter, and $L_p(t)$ using, for example, the half-space model or finite element models (Sato et al., 1990).

MA is usually characterized by very large deformations, with strains typically larger than 1. Drawbacks associated with this technique include the uncontrolled adhesion and friction between the glass micropipette and the cell, CSK detachment from the cell membrane during aspiration, and limited temporal resolution, especially at small times.

2.2 *Passive Techniques/Rheology*

2.2.1 One-Point and Two-Point Microrheology

One-point microrheology measures spontaneous displacement of a small probe, usually smaller than a micron, embedded in the cytoplasm of the living cell (Fig. 2a) (Mason, 2000; Mason and Weitz, 1995b; Tseng et al., 2002; Yamada et al., 2000). Thermal energy, k_BT, creates random force fluctuations on the probe, which, in turn, induce local deformations of the viscoelastic medium in the vicinity of the particle. From the measured mean square displacement, $\langle r^2(\Delta t)\rangle$, over time intervals, Δt, one can infer $G^*(f)$ using the Generalized Stokes-Einstein Relationship (GSER):

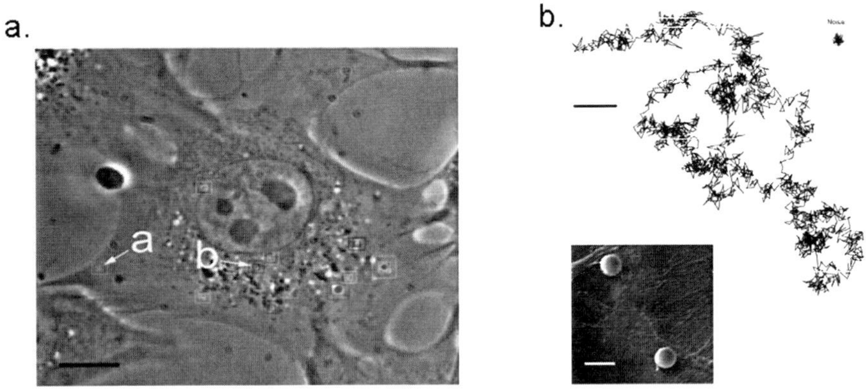

Fig. 2 Examples of passive techniques. (**a**) Multiple particle tracking of microspheres (~0.1 μm in diameter) injected in Swiss 3T3 fibroblast. Scale bar is 20 μm. Adapted from (Tseng et al., 2002). (**b**) Spontaneous trajectory of a 5 μm microbead attached to the apical surface of a HASM cell. Trajectory of bead glued to coverslip represents the upper limit of noise measurement. Black scale bar is 10 nm. Inset: EM of two RGD-coated microbeads. White scale bar is 5 μm. Adapted from (Bursac et al., 2005)

$$\langle r^2(\Delta t)\rangle = \Im^{-1}\left(\frac{k_B T}{\pi a i f G^*(f)}\right), \tag{1}$$

where a is the particle size, $\Im^{-1}(.)$ denotes the inverse Fourier Transform, f the frequency, and $i^2 = -1$. The GSER is the Stokes–Einstein relationship generalized to a frequency-dependent viscoelastic material; it is valid only for systems at thermodynamic equilibrium, in which case motions are driven by thermal forces alone.

Equation (1) is subject to the same conditions as the Stokes calculation: spherical tracer particle in a homogeneous, incompressible continuum with no–slip boundary conditions. But because heterogeneities of the medium and imperfect coupling of the probe to the surrounding medium are difficult to quantify, two-point microrheology (TPM) was recently developed (Crocker et al., 2000; Levine and Lubensky, 2001). TPM measures the cross-correlated motions of two separated tracers. The vector displacements of the tracers, $\Delta r_\alpha = r_\alpha(t+\Delta t) - r_\alpha(t)$, are first calculated, with t the absolute time, Δt the time interval, and α a spatial coordinate. Then the ensemble averaged tensor product of the tracer displacements is inferred: $D_{\alpha,\beta}(t,\Delta t) = < \Delta r^i_\alpha(t,\Delta t)\, \Delta r^j_\beta(t,\Delta t)\, \delta[r - R^{ij}(t)] >_{i\neq j,t}$, with α and β different spatial coordinates, $\delta(.)$ the Dirac delta function, and R^{ij} the distance between particles i and j. In the limit $r >> a$, the complex modulus, $\tilde{G}(s)$, in the Laplace domain is then given by:

$$\tilde{D}_{rr}(r,s) = \frac{k_B T}{2\pi r s \tilde{G}(s)}, \tag{2}$$

with s being the Laplace frequency. This method is independent of the tracer's size, shape, and the nature of the coupling between the tracer and the medium.

Like one-point microrheology, TPM also rests on the assumption of thermodynamic equilibrium and dynamics driven by thermal forces alone.

The major problem with the use of one-point and two-point microrheology in the living cell is that these techniques are based on the fluctuation dissipation theorem (FDT). By combining independent measurements of G^* and $\langle r^2(\Delta t)\rangle$, Bursac et al. have established that in the living cell the FDT breaks down for $\Delta t > 0.1$ s and, as a result, the GSER fails (see Section 3.4.2) (Bursac et al., 2005). This result was recently confirmed in reconstituted active cytoskeletal networks (Mizuno et al., 2007). Breakdown of the FDT implies that the forces driving CSK rearrangements are not solely of a thermal nature, and in particular, do not obey Boltzmann statistics. In the literature, one-point microrheology yield mechanical properties differing by orders of magnitude from active techniques (Tseng and Wirtz, 2001; Yamada et al., 2000). Regarding TPM, active forces are removed by ATP depletion (Hoffman et al., 2006; Van Citters et al., 2006). Although ATP depletion makes it technically possible to apply the GSER – and, therefore, TPM – the physiological relevance of such preparations becomes problematic.

2.2.2 Anomalous Bead Diffusion

Several laboratories including our own have established that the CSK is in a continuous state of turnover and is dramatically malleable (Gunst et al., 1995; Pratusevich et al., 1995; Smith et al., 2003). To measure cytoskeletal remodelling, we have recently developed a technique which consists of measuring spontaneous motions of an individual microbead anchored to the CSK (Fig. 2b) (An et al., 2004; Bursac et al., 2005; Bursac et al., 2007; Lenormand et al., 2007b). We hypothesize that the microbead can move only if the cytoskeletal structure to which it is attached rearranges. If so, then the measured mean square displacement (MSD) reports ongoing internal CSK reorganizations over time.

The MSD is defined as $\langle r^2(\Delta t)\rangle = \langle (r(t+\Delta t) - r(t))^2\rangle$ where r(t) is the bead position at time t, and Δt is the time lag. The brackets indicate either an average over all beads (An et al., 2004) or over t and all beads (Bursac et al., 2005). Experimental data show that the MSD increases as Δt^β. For Brownian motion, β is unity. If β is smaller than unity the motion is called subdiffusive, and when it is larger than unity the motion is called superdiffusive. When β is two, the motion is ballistic. Limitations and artefacts associated with this technique are the same as those associated with bead twisting (see Section 2.1.1).

2.3 Do These Techniques Measure CSK Dynamics?

2.3.1 Coupling a Functionalized Probe to the Cell Surface

The coupling of a functionalized probe to the cell surface is a complex issue. There are no independent means to confirm that the dynamics observed using functionalized surface-bound probes are solely a response of the CSK rather than of the probe-integrin receptor interaction or focal adhesion dynamics, and that this remains true at every frequency being probed. However, a compelling weight of evidence can be now found in the literature suggesting that the response obtained from functionalized microbeads is mainly dominated by CSK dynamics (Bursac et al., 2007, 2005; Fabry et al., 2003; Lenormand et al., 2007a,b; Puig-de-Morales et al., 2004; Trepat et al., 2007). When a microbead is coated with a peptide containing the sequence Arg-Gly-Asp (RGD), and incubated on the surface of a living cell, it forms focal adhesion complexes via integrin receptors. Mechanical responses of such a bead have been shown to be sensitive to manipulations of myosin motors, actin filaments, heat shock protein 27, filamin, intermediate filaments, cytoskeletal tension, depletion of ATP, and concentrations of a wide panels of contractile and relaxing agonists. At long time scale, RGD-coated beads display hopping motions that are different from a simple Brownian diffusion process in the membrane, and that correlate with the parameter x measured with active rheology (Fig. 8). Moreover, cholesterol depletion of the plasma membrane with cyclodextrin had no effect on CSK stiffness and spontaneous motions of an RGD-coated bead (Bursac et al., 2007), although changes in stiffness were observed using micropipette aspiration (Byfield et al., 2004).

Finally, in a recent study that used OMTC to probe the freshly isolated living ASM cell (Deng et al., 2006), Deng et al. demonstrated the existence of a high frequency regime in which the absolute magnitude and the frequency dependence of the complex modulus correspond precisely to entropic dynamics predicted for semi-flexible actin filaments driven by thermal forces (Gittes and MacKintosh, 1998). By establishing the hallmarks of entropic elasticity of a cross-linked actin network both qualitatively and quantitatively, these findings imply that, at high frequencies at least, the response of such a bead is not dominated by integrin dynamics, but rather by actin filaments.

2.3.2 Innate Variability and Detecting Physiologically Relevant Changes

We want to point out here the importance of obtaining measurements on a large number of cells in order to be able to report physiologically relevant responses. Mechanical properties of cells in culture are innately variable from cell to cell; Fabry et al. have shown that cell stiffness exhibits a log-normal distribution with a geometric standard deviation of almost 3 (Fabry et al., 2003, 2001b). Although the relative importance of potential sources of variability could not be quantified when this variability was first reported, it was attributed not only to innate variations in cell material properties but also variations between cell regions, and variations in the contact area between the bead and the cell. However, in measurements performed using USR, Desprat et al. reported a similar distribution (Desprat et al., 2005). Interestingly, in USR the entire cell is probed, the geometry of the deformation is controlled and fixed, and the contact area between the plate and the cell is taken into account. From these important observations we learned that the principal source of variability is innate variations in material properties rather than factors associated with the probe. This conclusion was subsequently verified using TPM (Hoffman et al., 2006; Van Citters et al., 2006).

Because the variability is so large, statistically relevant differences between groups of measurements can only be reported by acquiring a large amount of data – detecting differences of 25% using an unpaired, one-tailed t-test requires at least 50 cells per group (Fabry et al., 2001b). Although it imposes no difficulty for OMTC, which can probe hundreds of individual microbeads on hundreds of individual cells in the same culture dish simultaneously, for most techniques acquiring a relatively large number of measurements is problematic.

3 Properties of the Cytoskeleton

Although red blood cells (Henon et al., 1999; Puig-de-Morales-Marinkovic et al., 2007), cells with a more organised CSK such as cardiomyocytes (Lenormand G., unpublished data), and some cells studied in suspension (Wottawah et al., 2005) are notable exceptions, the behaviour of the CSK in a wide variety of living adherent cells is characterised by rather strange but very simple phenomenological laws. Where by ‘simple’ we mean that they capture the essence of the data with very few parameters even though the mechanistic basis is not presently understood, and by ‘law’ we mean a behaviour that is virtually universal over a wide range of cellular systems. These phenomenological laws are:

i. The CSK is soft and prestressed (Stamenovic et al., 2004; Wang et al., 2001).
ii. Although there are departures at very small time scales (Deng et al., 2006) and very large time scales (Stamenovic et al., 2007), CSK dynamics exhibits an extensive regime in which responses are scale-free (Fabry et al., 2001a, 2003).
iii. In response to stretch, the CSK promptly fluidizes and then slowly resolidifies (Bursac et al., 2005; Trepat et al., 2007).
iv. CSK rearrangements are intermittent and superdiffusive (Bursac et al., 2005).
v. All of the above are interconnected and biologically regulated.

3.1 *Law 1: The CSK is Soft and Prestressed*

The elastic modulus of the living cell is of the order $10^2 - 10^4$ Pa, showing that the cell is a very soft material indeed (Fabry et al., 2001a, 2003). Moreover the CSK is often described as a structure whose shape is stabilized predominantly by the tensile stresses borne by its contractile filaments (Ingber et al., 1994). Since its introduction, this hypothesis, known as the tensegrity hypothesis, has been controversial. The controversy was attributable mainly to a lack of experimental data that could be used to put the hypothesis to any sort of experimental test. However, the subsequent invention of OMTC and traction microscopy (Butler et al., 2002; Dembo and Wang, 1999), taken together with the advent of theoretical predictions (Stamenovic and Coughlin, 1999), opened tensegrity to rigorous experimental evaluation.

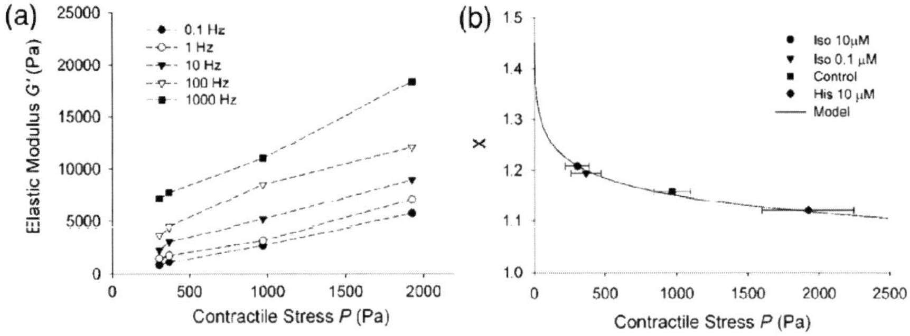

Fig. 3 **(a)** G′ increases with cytoskeletal contractile stress (or prestress) at all frequencies. **(b)** The parameter x (see text) decreases with increasing contractile stress. Adapted from (Stamenovic et al., 2004)

The tensegrity hypothesis predicts that cell stiffness must increase in proportion with the level of the tensile stress, which is also called the prestress. Using traction microscopy and OMTC, Wang et al. were able to put this hypothesis to its first experimental tests (Wang et al., 2001, 2002). Traction microscopy is used to measure the distribution of contractile stresses arising at the interface between each cell and its substrate; this distribution is called the traction field. Because the traction field must be balanced by tensile stresses within the cell body, the prestress could be measured directly, and cell stiffness, G′, measured by OMTC (Fig. 3a). As the contractile state of the cell is modulated with graded concentrations of relaxing or contracting agonists (isoproterenol or histamine, respectively), the prestress is modulated from 350 to 1,900 Pa. Over that range, cell stiffness increases linearly with the prestress, and is of the same order of magnitude, thus confirming two key *a priori* mechanistic, quantitative predictions. While this association does not necessarily preclude other interpretations, it is the hallmark of systems that secure shape stability mainly through prestress. Regardless of mechanism, these data establish a strong association between stiffness and the level of tensile stress within the CSK.

Finally, Stamenovic et al. reported a dependence of the prestress upon the parameter x, defined from the power law dependence on frequency f, of $G'(f) \sim f^{x-1}$ (Fig. 3b) (Stamenovic et al., 2004). As described below, the closer x is to 1, the closer is the system to ideally elastic behaviour; and similarly the closer x is to 2, the closer is the system to a Newtonian fluid. Although the physical basis remains to be explained, the dependence reported by Stamenovic et al. suggests the possibility that the CSK contractile stress may regulate transition between solid-like to fluid-like behaviour.

3.2 Law 2: Cytoskeleton Dynamics are Scale-Free

In a pivotal study that challenged long-held beliefs concerning cytoskeletal dynamics, Fabry et al. reported storage and loss moduli of the CSK of living adherent cells

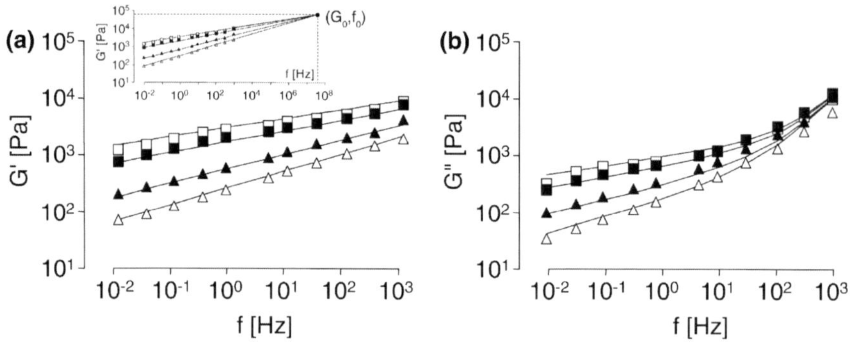

Fig. 4 Power law responses of G′ **(a)** and G″ **(b)** of HASM cells versus frequency under control conditions (■, n = 256), and after 10 min treatment with histamine [10^{-4} M] (□, n = 195), DBcAMP [10^{-3} M] (▲, n = 239) and cytochalasin D [2×10^{-6} M] (Δ, n = 171). Inset: Extrapolation of Eq. (3) to higher frequencies yields crossover at coordinate (G_0, f_0). Solid line: fit of the structural damping law. Adapted from (Fabry et al., 2001a). Reprinted figure with permission from Fabry, B., Maksym, G. N., Butler, J. P., Glogauer, M., Navajas, D. and Fredberg, J. J., Physical Review Letters, 87:148102, 2001. Copyright (2007) by the American Physical Society

using OMTC. Contrary to what had been previously believed (see Section 4.1), Fabry et al. established over a spectrum spanning five decades of frequency and in five different cell types that dynamics of the integrated cytoskeletal matrix cannot be tied to any characteristic time scale. Rather, relaxation times are distributed as a power law. The storage modulus, G′, increases only weakly with frequency, G′ (f) $\sim f^{x-1}$ with x = 1.18 (Fig. 4a), and the loss modulus, G″, follows a power law with the same exponent for frequencies below 30 Hz; at higher frequencies, a stronger frequency dependence progressively emerges (Fig. 4b). The loss tangent or hysteresivity, $\eta = G''/G'$, remains almost constant between 0.2 and 0.4 over the lower 4 frequency decades. Insofar as rheology is a window into underlying molecular dynamics, these results implied that no distinct internal molecular relaxation time scale could typify protein-protein interactions within the living CSK; such a response is called scale-free (Fabry et al., 2001a, 2003). Although the finding of scale-free responses was controversial when it was first reported, it has since been replicated by many investigators using a wide variety of methodologies (Alcaraz et al., 2003; Balland et al., 2006; Dahl et al., 2005; Desprat et al., 2005; Hoffman et al., 2006; Kasza et al., 2007; Lau et al.,, 2003; Lenormand et al., 2004; Overby et al., 2005; Valentine et al., 2005; Van Citters et al., 2006; Yamada et al., 2000; Yanai et al., 2004).

The identification of a very broad regime of scale-free (power law) rheology implies that within the cell body relaxations at all time scales are present simultaneously. Therefore, to speak of a "time constant" in the context of cytoskeletal rheology is an idea that has no physical meaning (see Section 4.1). This is not to say that other time scales do not come into play, however. Other characteristic times have been observed, but rather than corresponding to a molecular relaxation time or time constant, these times usually demark transitions between regimes of behaviour.

As described below (Section 3.4), for example, within the CSK of the adherent living cell there is a transition time from subdiffusive to superdiffusive regimes of $\langle r^2(\Delta t)\rangle$, a transition time for breakdown of the FDT, and a transition time from antipersistence to persistence of CSK remodelling (Bursac et al., 2005; Lenormand et al., 2007b). Also the additive Newtonian viscosity μ, which is several orders of magnitude higher than the viscosity of the liquid surrounding the cell (Fabry et al., 2001a, 2003; Lenormand et al., 2004), is a measure of the CSK properties and as such represents another distinct high-frequency regime. Furthermore, Stamenovic et al. have shown that cells displayed several power law regimes, one regime similar to that reported by Fabry et al. (Fabry et al., 2001a, 2003), and another regime at low frequencies ($10^{-3}-10^{-1}$ Hz); the emergence of this second power law regime, and the transition time between these regimes represent a characteristic time scale (Stamenovic et al., 2007). Finally Deng et al. have reported a distinct high-frequency regime with dynamics of different origin than that of the soft glassy behaviour; the transition from the glassy regime to the entropic regime represents a characteristic time scale as well (Deng et al., 2006). In the remainder of this section we restrict ourselves to the scale-free regime.

When cells are activated with a contractile agonist (histamine, 100 μM), G' increases but still exhibits a weak power law dependence on frequency; x falls slightly to 1.17 (Fig. 4). When cells are relaxed with N6,2-O-dibutyryladenosine3,5-cyclic monophosphate (DBcAMP, 1 mM), G' decreases, and x increases to 1.28. When the actin network of the cells is disrupted with cytochalasin D (2 μM), G' decreases further, and x increases to 1.33. Remarkably, G' data define a family of curves that, when extrapolated to high frequency, appear to intersect at a single value, (G_0, f_0) (Fig. 4a inset), and the complex modulus, $G^* = G' + iG'$, conforms closely to a phenomenological law known in the literature as structural damping (Fabry et al., 2001a, 2003):

$$G^*(f) = G_0\Gamma(2-x)\left(\frac{if}{f_0}\right)^{x-1} + i\mu f, \qquad (3)$$

with G_0 and f_0 scale factors for stiffness and frequency, μ an additive Newtonian viscosity, and $\Gamma(.)$ the gamma function. The first term on the right hand side of Eq. (3) describes a relationship between the exponent of the power law (x-1) and the transition from Hookean solid-like ($x = 1$) to Newtonian liquid-like ($x = 2$) behaviour. Thus, all changes in G' that occur in response to drug treatment are accounted for by changes of x alone (Fig. 4a inset); changes in G' are accounted for by changes in x and μ (Fabry et al., 2003).

This peculiar scale-free behaviour has been reported in different cell types using very diverse experimental techniques. Among the cell types are human and rat airway smooth muscle cells, human bronchial epithelial cells, mouse embryonic carcinoma cells (F9) cells, mouse macrophages (J774A.1), human neutrophils, human alveolar (A549) and bronchial (BEAS-2B) epithelial cells, human foetal lung fibroblasts, mice myoblasts (C2-7), and African green monkey kidney epithelium (TC7) cells (Alcaraz et al., 2003; Balland et al., 2005; Dahl et al., 2005;

Desprat et al., 2005; Fabry et al., 2001a; Lau et al., 2003; Lenormand et al., 2004; Yanai et al., 2004). Among experimental techniques are optical tweezers (Balland et al., 2005; Yanai et al., 2004), atomic force microscopy (Alcaraz et al., 2003), two-point microrheology (Lau et al., 2003), laser-tracking microrheology (Yamada et al., 2000), micropipette aspiration (Dahl et al., 2005), uniaxial stretching rheometer (Desprat et al., 2005). These responses are also established on cytoplasmic extracts using a conventional rheometer (Valentine et al., 2005). Moreover, power law behaviour persist when probing the cell through a variety of receptor systems, each serving as a different molecular window on cytoskeletal dynamics (Puig-de-Morales et al., 2004), and modulating cell rheology using a wide panel of drugs (Laudadio et al., 2005). Note that power law rheology has been observed using AFM and embedded probes (Alcaraz et al., 2003; Hoffman et al., 2006; Lau et al., 2003; Liu et al., 2006; Van Citters et al., 2006), and therefore scale-free rheology does not depend on specific surface receptor linkage. All of these studies, performed on a large number of different cell types, using a large panel of external probes as well as internal particles, demonstrate that power law behaviour and scale-free rheology are a common feature of cell mechanics.

Despite important differences in mechanical properties, measurements performed on different cell types and after a wide variety of treatments collapse onto two simple relationships representing the elastic and frictional behaviours (Figs. 5a,b respectively) (Fabry et al., 2001a, 2003; Laudadio et al., 2005; Lenormand et al., 2004). The first relationship is obtained plotting a normalized stiffness, G, as G′ measured at 0.75 Hz (an arbitrary choice) divided by G_0 (defined in Eq. (3)) versus x. The second relationship is obtained plotting the hysteresivity η (measured as the ratio of

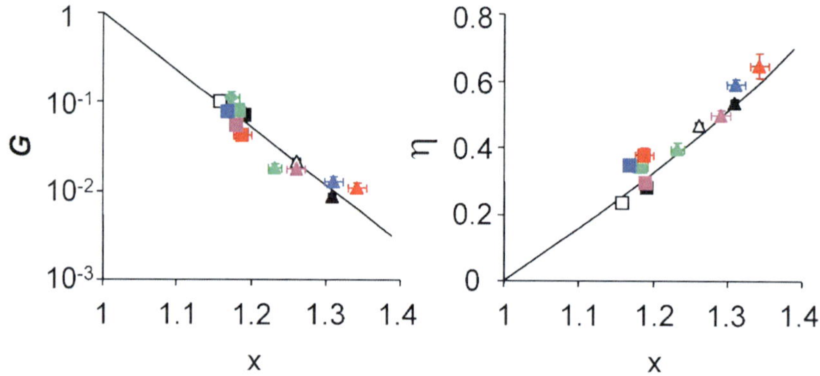

Fig. 5 Master curves (G and η at 0.75 Hz versus x) of HASM cells (*black*, n = 256), human bronchial epithelial cells (*blue*, n = 142), mouse embryonic carcinoma cells (F9) cells (*pink*, n = 50), mouse macrophages (J774A.1) (*red*, n = 46) and human neutrophils (*green*, n = 42) under control conditions (■), treatment with histamine (□), N-formyl-methionyl-leucyl phenylalanine (♦), DBcAMP (△), and cytochalasin D (▲). Eq. (3) predicts the *black solid* curves. Error bars indicate ± one standard error. From (Fabry et al., 2001a). Reprinted figure with permission from Fabry, B., Maksym, G. N., Butler, J. P., Glogauer, M., Navajas, D. and Fredberg, J. J., Physical Review Letters, 87:148102, 2001. Copyright (2007) by the American Physical Society

G''/G' at 0.75 Hz) versus x. These relationships represent universal master curves in that a single free parameter, x, defines the constitutive elastic and frictional behaviours for a variety of cytoskeletal manipulations, and for diverse cell types. These data suggest that the parameter x plays a role akin to a temperature: in the limit that x approaches 1, the system behaves as an ideal Hookean elastic solid, and in the limit that x approaches 2 the system behaves as an ideal Newtonian fluid. The living cell modulates its mechanical properties by moving between states that are 'cold,' frozen, and solid-like, and states that are 'hot,' melted, and liquid-like.

3.3 Law 3: Prompt Shear-Induced Fluidization Followed by Slow Resolidification

Large mechanical shear causes the CSK matrix to fluidize, a phenomenon similar to physical rejuvenation in SGMs. This fluidization is followed by a slow scale-free recovery of mechanical properties, a phenomenon similar to physical aging in SGMs. Surprisingly in response to a transient stretch, the CSK fluidizes in a pattern that is universal for different cell types (Trepat et al., 2007). This finding implicates mechanisms mediated not so much by specific signalling pathways, as it is usually assumed, but rather – as we explain below – by non-specific actions of physical forces.

3.3.1 Physical Rejuvenation and Physical Aging

We investigated aging and rejuvenation at the cellular level using two different approaches (Bursac et al., 2005; Lenormand et al., 2007a; Trepat et al., 2007). In Bursac et al. and Lenormand et al., a large oscillatory shear is produced locally by applying an oscillatory torque to the RGD-coated beads, whereas in Trepat et al., a transient uniform biaxial stretch is achieved globally by deforming a flexible membrane upon which the living cells are firmly adhering. In both methods, changes in mechanical properties are determined using our OMTC device. In both cases, mechanical properties become a function of the waiting time after cessation of the large stretch (Fig. 6a).

In Bursac et al. and Lenormand et al., rejuvenation is quantified as the creep function $J(t, t_w)$ measured at a waiting time of 10 s after the large shear cessation, expressed as a percent change of the creep function before shear; both creep functions are taken at $t = 1$ s (an arbitrary value) (Bursac et al., 2005; Lenormand et al., 2007a). When no shear is applied, no change in the creep function is observed, and rejuvenation is zero (Fig. 6b). As the amplitude of the oscillatory torque is progressively increased, creep measured upon cessation of the oscillatory torque increases as well. After applying a large shear, the CSK matrix becomes softer, indicating physical rejuvenation. Interestingly, when creep functions are measured at different waiting times, t_w, following the large oscillatory torque, $J(t, t_w)$ decreases systematically as t_w increases, showing that the system becomes progressively stiffer (Fig. 6c). Furthermore, for

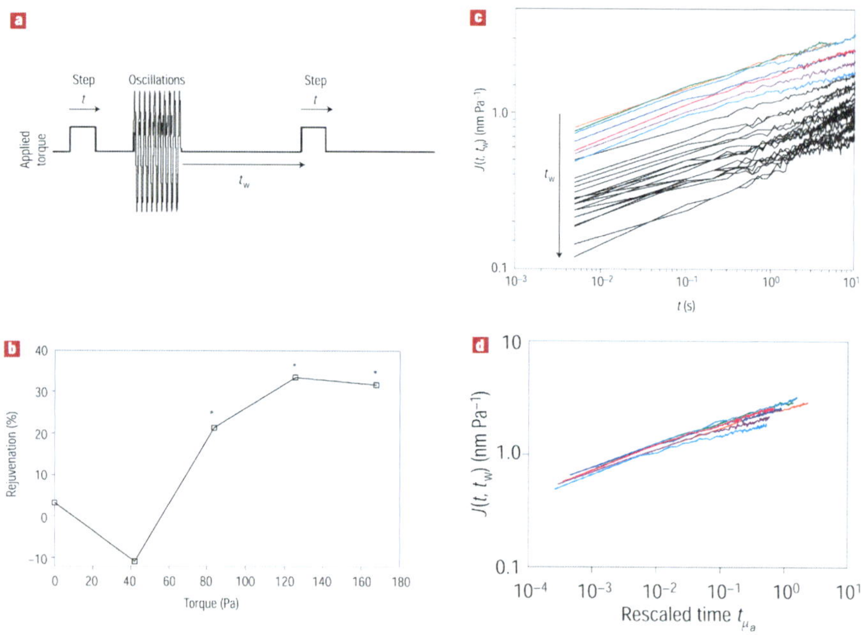

Fig. 6 Aging and rejuvenation in response to applied shear. **(a)** Applied torque history. **(b)** Rejuvenation is quantified as the relative per cent change between $J(t, t_w)$ at $t = 1$ s, $t_w = 10$ s after shear, and the creep at $t = 1$ s before shear (*: significant changes between the two creep measurements, $p < 0.001$). **(c)** Creep functions at different t_w (each line is median of over 658 beads on a like number of cells): 10 s (*red*), 40 s (*green*), 100 s (*dark cyan*), 400 s (*blue*), 800 s (*pink*), 1200 s (*dark pink*), 1600 s (*cyan*) and every 400 s up to 8000 s (*black*). The geometric standard deviation for each $J(t, t_w)$ ranged between 1.2 and 1.5. **(d)** For $t_w \leq 1600$ s, $J(t,t_w)$ collapsed onto a master curve using the rescaled time with $\mu_a = 0.4$. Adapted from (Bursac et al., 2005)

t_w below 1600 s, all creep functions collapse onto a unifying master curve using a rescaled time $t_{\mu_a} = t/t_w^{\mu_a}$ with the aging coefficient $\mu_a = 0.4$ (Fig. 6d). Slow evolution of rheological properties and collapse of this kind are typical of soft glassy systems that are aging (Cloitre et al., 2000; Derec et al., 2003; Ramos and Cipelletti, 2001).

In Trepat et al., the adherent living cells are stretched globally by a 4 s duration transient stretch (Trepat et al., 2007). Stiffness after stretch relative to stiffness of the same cell immediately before is denoted G'_n. When no stretch is applied, G'_n does not change. But immediately after cessation of a transient stretch, G'_n promptly decreases and then slowly recovers (Fig. 7a and b). Furthermore, immediately after stretch cessation, the phase angle $\delta = \tan^{-1}(G''/G')$ promptly increases and then slowly recovers (Fig. 7a and b, insets). Note that, at low frequencies, the phase angle is directly linked to the parameter x by the relation $\delta = \frac{\pi}{2}(x - 1)$. To assess the robustness of these responses, cells were pre-treated with an extensive set of mechanistically distinct drugs. Across the substantial panel of interventions, fluidization–resolidification responses to stretch are similar (Fig. 7a and b). These prompt changes establish that shear tends to fluidize the cell, comparable to the effect of shear on soft glassy materials (Mason and Weitz, 1995a; Sollich et al., 1997; Sollich, 1998).

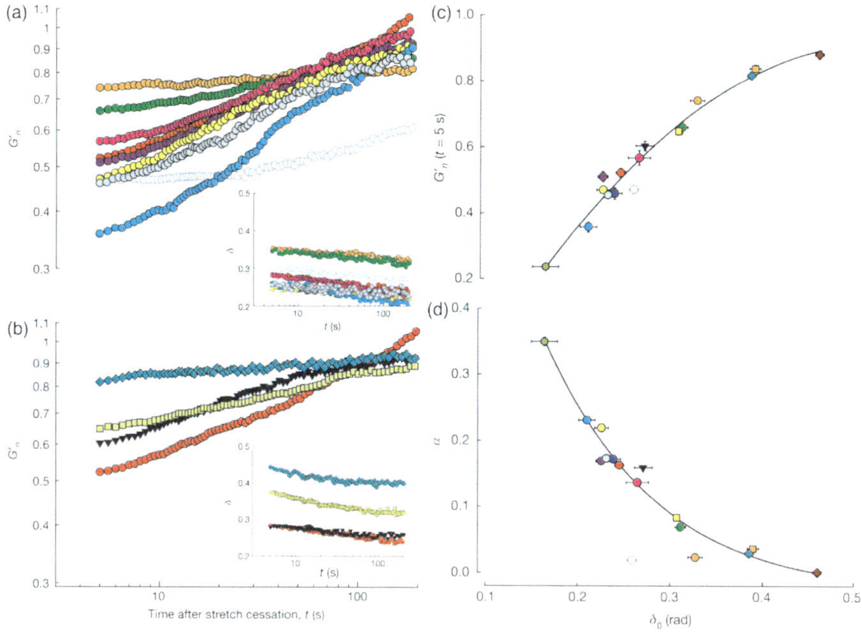

Fig. 7 A broad variety of cell systems are fluidized by a transient stretch of 10% amplitude. (**a**) G'_n and δ (inset) of pharmacologically treated HASM cells after application of a single transient stretch of 10% amplitude. Groups are latrunculin A (*orange*), DBcAMP (*green*), ML7 (*dark pink*), histamine (*yellow*), EGTA (*grey*), jasplakinolide (*bright blue*), ATP depletion (open symbols), and untreated cells (*red*). (**b**) G'_n and δ (inset) of Madin-Darby canine kidney (MDCK) epithelial (*blue diamonds*), human bronchial epithelial (HBE, *yellow squares*), human lung fibroblast (*black triangles*) and HASM (*red circles*). (**c**) Master curves of G'_n at the earliest time point recorded after stretch and of the initial rate of stiffness recovery α versus the pre-stretch phase angle δ_0 (**b**). α was assessed by fitting a power law $G'_n \propto t^\alpha$ to the first 30 s of response after stretch cessation. Error bars indicate standard errors. Colors are as above, with the addition of HASM in phosphate-buffered saline (*dark blue*), HBE Latrunculin A (*orange squares*), MDCK cytochalasin D (*brown diamonds*) and BASM tissue (*green hexagons*). Adapted from (Trepat et al., 2007)

3.3.2 Universal Physical Response to Stretch of the Adherent Living Cell

Focusing on the prompt stiffness reduction G'_n (assessed at the earliest measurable time point, t = 5 s) and its initial rate of recovery α (assessed from the fit of G'_n to t^α for the first 30 s after a sudden stretch), Trepat et al. established a striking unification of these diverse responses. Despite the broad diversity of drug interventions and cellular systems, all data collapsed onto a single unifying relationship when G'_n (at t = 5 s) was plotted against the pre-stretch value of the phase angle, δ_0 (Fig. 7c). Similarly, when the rate of stiffness recovery was plotted against δ_0 another master relationship was defined, although ATP depletion fell off that relationship (Fig. 7d). These two unifying relationships describe the response to stretch of a broad variety of cell systems. In every case, the closer the system is to the solid-like state ($\delta_0 = 0$)

before being subjected to transient stretch, the greater is the extent of its fluidization and, except for the case of ATP depletion, the faster is its subsequent recovery.

3.4 Law 4: CSK Rearrangements are Intermittent and Superdiffusive

As described in Section 2.2.2, we assess cytoskeletal rearrangements from spontaneous motions of microbeads tightly bound to the CSK of HASM cells. We reason that the microbead could move only if the structure to which it is attached rearranged.

3.4.1 Intermittent Dynamics of Spontaneous Structural Rearrangements

Spontaneous bead motions show intermittent dynamics, with periods of confinement (stalling) alternating with periods of persistent motion (hopping) (Fig. 2b). The mean square displacement, $\langle r^2(\Delta t)\rangle$, exhibits subdiffusive behaviour for small time lags, and superdiffusive behaviour for large time lags (Fig. 8a). The subdiffusive regime is associated with stalling, and the superdiffusive regime at larger times is associated with hopping.

The probability distribution of bead displacements shows non-Gaussian behaviour with broad tails attributable to transitions between stalling and hopping motions (Fig. 8a inset). The extent to which these distributions depart from Gaussian behaviour is quantified through the non-Gaussian parameter, α; α is zero for a Gaussian distribution and greater than zero when the distribution has broad non-Gaussian tails. α is calculated as the normalized kurtosis $\alpha = \langle\Omega\rangle^4/3\langle\Omega^2\rangle^2 - 1$, where Ω represents each bead displacement Δy normalized to its averaged absolute value, $\Omega = \Delta y/\left|\overline{\Delta y}\right|$; brackets $\langle.\rangle$ denote average over all beads in an experimental group. We define Δt as the time lag at which α reaches its peak value $\alpha(\Delta t^*)$ (Fig. 8c). Δt^* is found to correspond to the transition between subdiffusive and superdiffusive behaviour (Fig. 8a and c); the larger is Δt^*, the slower are system dynamics and the later is the onset of the superdiffusive regime. Intermittent dynamics associated with non-Gaussian statistics are signatures of glassy systems (Bissig et al., 2003; Cipelletti et al., 2003; Weeks et al., 2000), where slowing of dynamics and more prominent intermittency indicate approach of a glass transition and kinetic arrest (Kob et al., 1997; Weeks et al., 2000).

As described in Section 3.2, the cell storage modulus $G'(f)$ shows a weak power law dependence on frequency f, $G' \sim f^{x-1}$. The closer the value of x to 1, the closer is the system to ideally elastic behaviour. As the measured value of x approached 1, $\alpha(\Delta t^*)$ increases strongly (Fig. 8b), a behaviour reminiscent of colloidal systems approaching a glass transition, and indicative of approach to kinetic arrest (Kob et al., 1997; Weeks et al., 2000). However, Δt^* shows a weaker dependence on x, suggesting that the rate of CSK dynamics is not determined solely by x (Fig. 8b inset). Indeed, this point is underscored by closer examination of $\langle r^2(\Delta t)\rangle$ at long Δt; we fit the superdiffusive behaviour of $\langle r^2(\Delta t)\rangle$ to $D^*(\Delta t/t_0)^\beta$, where t_0 is arbitrarily

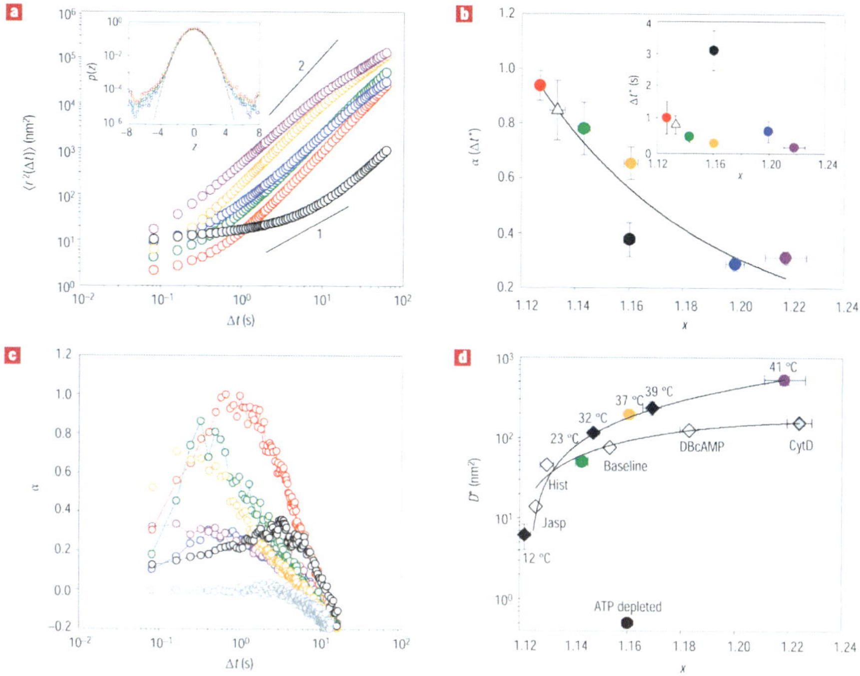

Fig. 8 Statistics of spontaneous bead motions. **(a)** Mean square displacement $\langle r^2(\Delta t)\rangle$($n = 400$–$720$ beads per group) versus time lag Δt. Groups are 23 °C (*green*), 37 °C (*orange*) and 41 °C (*purple*), DBcAMP (*blue*), Jasplakinolide (*red*), and ATP depletion (*black*). Solid lines indicate logarithmic slopes of 1 and 2. Inset: Probability density distribution p(z) of one-dimensional normalized bead displacements z, shown for time lags Δt^*. Line is best fit Gaussian. **(b)** $\alpha(\Delta t^*)$ decreases with increase in x. Open triangles: cells in PBS plus glucose (10 mM) at 23 °C. Inset: Δt^* versus x (bars in x-direction are standard errors; bars in y-direction are estimated from **(c)**. Solid line is a guide to the eye. **(c)** Non-Gaussian parameter α versus Δt. Colour coding the same as above, grey symbols represent beads glued to the coverslip. **(d)** Cytoskeletal rearrangements D^* increase with x for different thermodynamic temperatures or after CSK manipulations. Baseline condition measured at room temperature before adding agonists (Baseline), histamine (Hist), jasplakinolide (Jasp), and cytochalasin D (CytD). Solid lines are a guide to the eye. From (Bursac et al., 2005)

taken to be 1 s (Fig. 8d). The larger is x, the larger is D^*, but these responses do not show a unique relationship with x; changes induced by variations in thermodynamic temperature follow a different relationship than those induced by drugs that modulate CSK structures, or those induced by depletion of intracellular ATP. These observations suggest that CSK rearrangements in the long time limit are strongly influenced by mechanical energy released during ATP utilization.

3.4.2 Break-Down of FDT and the Generalized Stokes Einstein Relationship (GSER)

We now discuss the extent to which the anomalous motions described above might be accounted for by force fluctuations of thermal origin (Bursac et al., 2005). Bursac et al. calculated $\langle r^2(\Delta t)\rangle$ from measurements of $G^*(f)$ using the GSER (see Eq. (1),

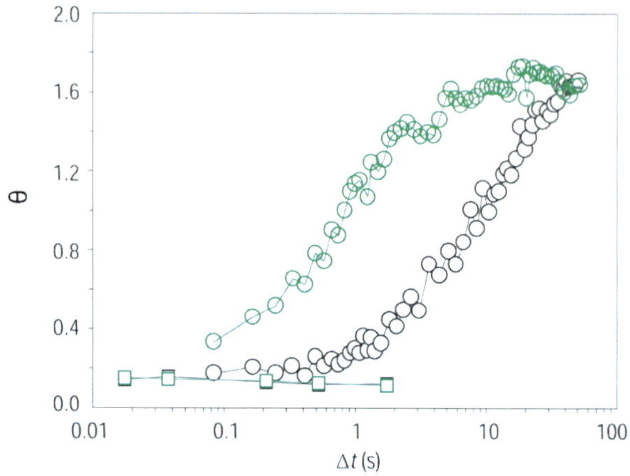

Fig. 9 Breakdown of the generalized Stokes Einstein relationship. Local slope θ (i.e., $\langle r^2(\Delta t)\rangle \sim \Delta t^{\theta}$) determined from direct observations of $\langle r^2(\Delta t)\rangle$ (*circles*) and from predictions based on GSER and measurements of $G^*(f)$ (*squares*), under baseline condition (green) and after ATP depletion (black). Adapted from (Bursac et al., 2005)

Section 2.2.1). They focused on the local exponent θ (i.e., $\langle r^2(\Delta t)\rangle \sim \Delta t^{\theta}$), and compared these slopes with those measured directly from spontaneous bead motions (as in Fig. 8a). Measurements approach the predictions from GSER for the smallest measured time lags (Fig. 9). But for longer times, predictions and data diverge strongly, indicating a clear breakdown of GSER. The breakdown of the GSER implies that the system is away from thermodynamic equilibrium, and that mechanical energy dissipation has become uncoupled from force fluctuations of thermal origin. When ATP in the cell is largely depleted, a closer correspondence between data and predictions is observed, however, and breakdown of the GSER is substantially delayed (Fig. 9).

3.5 Law 5: Laws 1-4 are Interconnected and Biologically Regulated

We have shown above that the parameter x is directly linked to the cytoskeletal prestress (Law 1, Fig. 3b). The parameter x controls the CSK's position along a continuous spectrum of solid-like versus and fluid-like states (Law 2, Fig. 5). This very same parameter x predicts shear-softening and the rate of recovery (Law 3, Fig. 7c and d). Although it does not show an entirely unique relationship, the parameter x is the main factor controlling the rate of cytoskeletal structural reorganization and the intermittent non-Gaussian nature of its statistics (Law 4, Fig. 8d). These observations, taken together, lead to the 5th law, which is that a cell modulates its mechanical behaviour mainly by modifying the sole parameter x. Biological interventions seem to modify CSK dynamics only to the extent that they affect x. Although x is rather easy to measure, what it means mechanistically is quite a different matter. As described below, important clues may come from an unexpected and apparently unconnected field, namely soft condensed matter physics.

4 The CSK as a Soft Glassy Material Close to the Glass Transition

4.1 Why do Classical Viscoelastic Models not Work?

It is important to recognize that the power law mechanical behaviour reported here (Section 3.2) stands in contrast with the prevailing paradigm of cell mechanics, which holds that cell mechanical properties arise from a small number of distinct elastic and viscous components, and therefore express only a limited range of characteristic relaxation times. To illustrate this disparity, we fit a creep response (Lenormand et al., 2004) with the Burgers model (Findley et al., 1989) (essentially equivalent to that used by Bausch et al. (Bausch et al., 1998)). The difference between the fit of the Burgers model and that of a power law is hardly detectable when data span only a single time decade (Fig. 10a), however, over a wider time window, the Burgers model is clearly unable to account for the creep response despite the fact that it has four free parameters (Fig. 10b). Furthermore, in the Burger model, elasticity and viscosity arise from distinct structures. However, our data show (Section 3.2) that for frequency below 30 Hz, G' and G'' follow a weak power law with the same exponent, and the ratio G''/G' is constant and in the order of 0.1. This implies that, within the CSK, dissipation is intrinsically linked to elasticity.

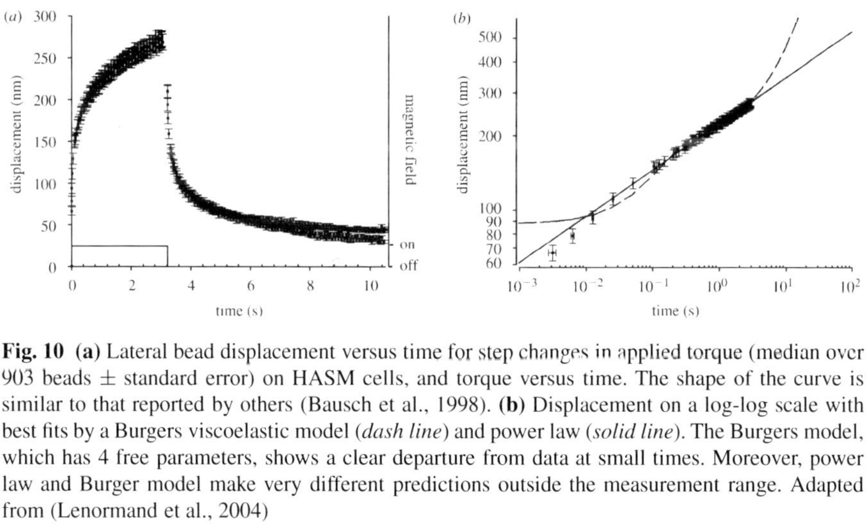

Fig. 10 **(a)** Lateral bead displacement versus time for step changes in applied torque (median over 903 beads ± standard error) on HASM cells, and torque versus time. The shape of the curve is similar to that reported by others (Bausch et al., 1998). **(b)** Displacement on a log-log scale with best fits by a Burgers viscoelastic model (*dash line*) and power law (*solid line*). The Burgers model, which has 4 free parameters, shows a clear departure from data at small times. Moreover, power law and Burger model make very different predictions outside the measurement range. Adapted from (Lenormand et al., 2004)

4.2 What are Soft Glassy Materials (SGMs)?

The term 'soft matter' is a recent addition to the physics vocabulary. Pierre-Gilles de Gennes was one of the first to use it in his Nobel lecture in 1991. 'Soft matter' covers colloids, polymers, surfactant phases, emulsions, and granular media. Though very different in structure, the mechanical behaviour of each of these substances is surprisingly alike. All are very soft (in the range of Pa to kPa), both the storage, G′, and the loss, G″, moduli increase with the same weak power law dependences on frequency, and the ratio G″/G′ is frequency insensitive and in the order of 0.1 (Sollich, 1998). The fact that similar anomalous rheology should be seen in such a wide range of soft materials suggests a common cause (Sollich et al., 1997). Sollich et al. argued that 'glassy' dynamics is a natural consequence of two properties shared by all soft materials: structural disorder and metastability.

4.3 Sollich's Soft Glassy Rheology (SGR) Model

Sollich developed a theory of soft glassy rheology (SGR) using an earlier work by Bouchaud (Bouchaud, 1992; Sollich, 1998; Sollich et al., 1997). The SGR model is a phenomenological model that aims to explain the main features of SGM rheology. To be reasonably generic, it incorporates only a minimal number of features common to all SGMs, leaving aside as much system specific details as possible. One important feature is the 'glassiness' i.e., the effects of structural disorder and metastability. This feature is modelled using a fairly intuitive picture of a glass (Fig. 11): it consists of local 'elements' that are trapped in 'cages' formed by their neighbours so that they cannot move. Occasionally, however, a rearrangement of the elements may be possible, due to thermal activation, for example. In SGMs however, it is unlikely that thermal motion alone is enough to archive complete structural relaxation. It is assumed instead that the 'activation' in SGMs is due to 'interactions': a rearrangement somewhere in the material can propagate and cause

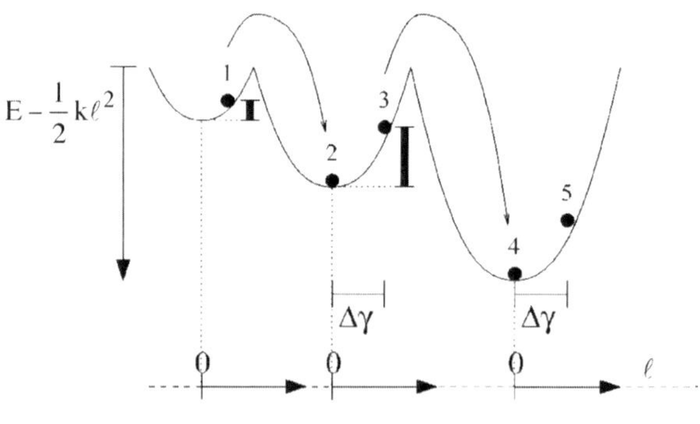

Fig. 11 Potential well picture of the dynamics of the SGR model. Note that the relative horizontal displacement of the quadratic potential wells is arbitrary; each has its own independent zero for the scale of the local strain ℓ. The solid vertical bars indicate the energy dissipated in the 'hops' (yield events) from 1 to 2 and 3 to 4, respectively. When a particle hops, the stored elastic energy $E = \frac{1}{2}k\ell^2$ dissipates, giving rise to the link between G′ and G″. Adapted from (Sollich, 1998). Reprinted figure with permission from Sollich P., Physical Review E, 58, 738–759, 1998). Copyright (2007) by the American Physical Society

rearrangements elsewhere. In a mean-field spirit, this coupling between elements is represented by an effective temperature x.

To describe rheology and flow, the SGR incorporates a local strain variable ℓ. As the system is sheared, a constitutive element first deform elastically from a local equilibrium configuration, giving rise to a stored elastic energy (Fig. 11). This continues up to a yield point, ℓ_y, whereupon rearrangement occurs. The yield energy is $E = \frac{1}{2}k\ell_y{}^2$, where k is an elastic constant. The effects of structural disorders are modelled by assuming a distribution of yield energies E, rather than a single value common to all elements. While the SGR model assumes a spatially homogeneous strain rate, it does admit inhomogeneities in the local strain ℓ and stress $\sigma = k\ell$. These inhomogeneities arise because different elements generally yield at different times. Therefore to describe the state of the system at a given time, one now needs to know the joint probability $P(E, \ell, t)$ of finding an element with a yield energy E and a local strain ℓ at time t. Within the SGR model this probability evolves in time according to:

$$\partial_t P = -\dot{\gamma}\partial_\ell P - \Gamma_0 P e^{-[E-k\ell^2]/x} + \Gamma(t)\rho(E)\delta(\ell), \tag{4}$$

where

$$\Gamma(\mathrm{t}) = \Gamma_0 \int \mathrm{dE}\, \mathrm{d}\ell\, \mathrm{P}(\mathrm{E}, \ell, \mathrm{t})\mathrm{e}^{-[\mathrm{E}-\mathrm{k}\ell^2/2]/\mathrm{x}}, \tag{5}$$

Γ_0 is the transition rate constant, and x is the effective temperature. The macroscopic stress is given by the average of the local stress, $\sigma(E) = k \int \ell P(E, \ell, t) dE d\ell$. These three equations, taken with the distribution of trap depths, $\rho(\mathrm{E}) \sim \mathrm{e}^{-\mathrm{E}}$, define the SGR model.

4.4 SGR Applied to the Living Cell

Remarkably, in the limit that the frequency is small compare to Γ_0, the SGR model leads directly to the structural damping equation (Eq. 3). If Sollich's theory is applied to the CSK, then the parameter x corresponds to the noise temperature of the cytoskeletal matrix, and represent the amount of 'jostling' within the CSK relative to the depth of energy wells, G_0 is the CSK stiffness at the glass transition, and f_0 corresponds to Γ_0, and is the maximum rate at which cytoskeletal elements can escape their traps. It is important to note that SGR does not invoke a viscosity. Energy dissipation in SGR theory comes not at all from a viscous process but from elastic strain energy that is stored, and, upon a hopping transition, irreversibly lost. Finally note that when the CSK is modulated using specific drugs, we hypothesize that changes arise from modulation of x, and that we assume the glass transition temperature, x_g, to remain constant.

A physical picture capturing these dynamics uses an energy landscape to describe all possible configurations of the CSK. A cytoskeletal element (e.g., structural protein or protein complex) finds itself trapped in an energy well. Trapping might arise from physical interactions such as molecular crowding, attractive interactions, or weak chemical bonding. Thermal energy is typically insufficient to push the system over energy barriers into different configurations; this is analogous to a jammed state (Liu and Nagel, 1998; Trappe et al., 2001). Through intermittent transitions from one metastable state to another, the system evolves slowly (ages) into more stable configurations. Imposed mechanical stress can provide sufficient energy to bring the system to a new configuration, thus pushing the system farther from thermodynamic equilibrium and resetting this evolution (rejuvenation). Applied to the living cell, this hypothesis implies that molecular crowding and approach to kinetic arrest are key attributes of CSK dynamics.

5 Open Questions

5.1 What is the Role of ATP?

Sollich interprets the effective temperature x as reflecting jostling of elements (by an unidentified but non-thermal mechanism) relative to the depth of energy wells. It appears an interesting question whether the ambiguity surrounding x might be resolved in the case of the living cell where the mechanical energy released by ATP hydrolysis is quite large. In the living cell, the energy derived from hydrolysis of the gamma phosphate bond of ATP is roughly 25 k_BT per event (Howard, 2001); and for one cubic micron of cytoplasm of a typical cell, these events occur at a rate of more than 10^4/s (Gunst and Fredberg, 2003). Vesicle transport and any local motions caused by proteins that undergo cyclic conformational changes, such as molecular motors, contribute to the intracellular molecular agitation. Associated local motions are quite large, in the range 1–10 nm, which is comparable to molecular dimensions and molecular spacing.

ATP-dependent rearrangements of the CSK modify the configurations themselves, providing an alternate means of exploring new configurations; ATP hydrolysis can drive both protein conformational changes and polymerization/ depolymerization cycles, either of which could conceivably resolve constraints and drive structural rearrangements. This is radically different from traditional glassy systems which remain trapped in a configuration for increasingly long times because only thermal energy is available to allow the system to evolve. Interestingly, x does not seem to be affected by ATP depletion (Fig. 8d and (Bursac et al., 2005)). But at the same time, it is possible that ATP also acts to tilt the energy landscape in such a way as to decrease energy barriers.

At the level of physics, whether E is an energy is a question of much debate. In Sollich's theory, $<E>$ (the ensemble average of E over the probability density of E) is not a conserved quantity. This is not a fatal flaw in non-isolated systems which are in weak contact with an external heat reservoir where $<E>$ can either increase or decrease or stay the same, depending on the initial conditions of the prepared state. This same idea is true in the Sollich's theory – the evolution of <E> mirrors that of the evolution of the probability density of E. If the initial state is prepared in a high <E> configuration (perhaps associated with fluidization), then <E> will gradually decrease, and conversely, if prepared in a low <E> state, than <E> will gradually increase, both behaviors tending toward the steady state <E>, and with an approach that is temporally a power law.

5.2 What is the Meaning of the Common Intersection?

The existence of a common intersection at high frequency, f_0, remains an empirical finding with no known mechanistic basis. Nonetheless it is highly consequential because it accounts for the fact that all G' data can be described by only one parameter, x, and therefore collapse onto unifying master relationships. Statistical analysis of our data suggests that f_0 did not vary with drug treatments and possibly not even across cell type (Fabry et al., 2003). In Sollich's theory, f_0 is identified as being the maximum rate at which cytoskeletal elements can escape their traps. However, for soft glassy materials in general, and for the living cells in particular, the factors that determine f_0 remain unclear, and why f_0 is invariant is not explained by SGR theory.

Factors that are present in all cells are the presence of water and molecular crowding, with an average distance between macromolecules of only 2 nm, or roughly 10 water molecules across. Using microwave dielectric spectroscopy, Suzuki et al. have measured the rotational mobility of water in the vicinity of proteins, and they obtained a relaxation frequency ranging from a few GHz to 40 GHz (Fuller and Rand, 1999; Kabir et al., 2003; Suzuki et al., 2004). This range of relaxation frequencies coincides with the range of estimates for the intersection frequency f_0. We speculate that if the cell were perturbed mechanically at a frequency corresponding to the rotational relaxation frequency of water, the water might dominate the mechanical response of the system and differences in stiffness with various CSK interventions might vanish. Whether this mechanism accounts for the observed intersection remains a matter of speculation.

Interestingly, Kramers-Smoluchowski theory of the overdamped unbinding kinetics of weakly bonded structures involves a diffusive flux over an energy barrier, with an attempt frequency $\Phi_0 = D/l_c l_{ts}$, where D is the diffusion constant related to molecular damping, and the length scales l_c and l_{ts} are the thermal widths of the bound states and of the barrier respectively (Evans, 2001). This theory is used to describe force spectroscopy of single molecular bonds, and typically gives Φ_0 in the order of 10^9–10^{10} Hz, which is consistent with the value of f_0 extrapolated from the measurements of $G^*(f)$ (Fabry et al., 2001a, 2003; Lenormand et al., 2004; Puig-de-Morales et al., 2004). The striking similarity between f_0 and Φ_0 could mean that scale-free rheology is a consequence of proteins unfolding under forces. This point is further discussed in the following section.

5.3 Protein Unfolding and Glassy Rheology

Glassy dynamics come into play at the level of forced unfolding of isolated protein domains (Brujic et al., 2006). Using single-molecule force-clamp spectroscopy to measure the kinetics of unfolding of the isolated ubiquitin molecule, Brujic et al. have shown that the distribution of unfolding rates is surprisingly broad and follows a power law with no characteristic mean. Structural relaxations are governed by an exponential distribution of energy well depths with a distribution of 5–10 $k_B T$. Corresponding interconversion between conformations is scale-free and fails to conform to the modified Arrhenius form. Accordingly, conformations may unfold only at sufficiently high forces that barriers become comparable to $k_B T$ but otherwise remain frozen into a given distribution and away from thermodynamic equilibrium.

5.4 What Can We Learn from Reconstituted F-Actin Networks?

In the past few years, reconstituted cytoskeletal protein networks have been shown to mimic many aspects of the mechanical properties of the cell, providing new insight into the origin of cellular behaviours (Kasza et al., 2007). Filament length and crosslinking protein concentration in networks of F-actin are among the most widely studied (Gardel et al., 2004; Liu et al., 2006). Interestingly, with increasing applied stress, these networks stiffen. This stress-stiffening has important implications: the magnitude of their linear elasticity is typically orders of magnitude less than that of cells, but when prestressed into the nonlinear regime the elasticity of these networks dramatically increases and approaches that of cells (Gardel et al., 2006). Recently, a reversible stress-softening behaviour in a reconstituted actin network has been reported (Chaudhuri et al., 2007). These reconstituted networks provide ideal systems to study the complex mechanical behaviours of the living cell, and in particular the interaction between prestress and power law rheology (Stamenovic, 2006).

6 Concluding Remarks

The CSK of the living cell exhibits physical properties and remodelling dynamics with all the same signatures as inert condensed systems, although with important differences as well. In these systems, the transition between solid-like and liquid-like states is not at all understood, and represents an outstanding unsolved problem in the field of soft condensed matter. Framing CSK dynamics in this peculiar context brings trapping, intermittency, and approach to kinetic arrest as key features of collective protein–protein interactions. These generic features represent novel mesoscale mechanisms that might link integrative CSK functions to underlying molecular events.

References

A-Hassan, E., Heinz, W. F., Antonik, M. D., D'Costa, N. P., Nageswaran, S., Schoenenberger, C. A. and Hoh, J. H., 1998, Relative microelastic mapping of living cells by atomic force microscopy, Biophys. J. 74, 1564–78.

Alcaraz, J., Buscemi, L., Grabulosa, M., Trepat, X., Fabry, B., Farre, R. and Navajas, D., 2003, Microrheology of human lung epithelial cells measured by atomic force microscopy, Biophys. J. 84, 2071–9.

Alcaraz, J., Buscemi, L., Puig-De-Morales, M., Colchero, J., Baro, A. M. and Navajas, D., 2002, Correction of microrheological measurements of soft samples with atomic force microscopy for the hydrodynamic drag on the cantilever, Langmuir 18, 716–21.

An, S. S., Fabry, B., Mellema, M., Bursac, P., Gerthoffer, W. T., Kayyali, U. S., Gaestel, M., Shore, S. A. and Fredberg, J. J., 2004, Role of heat shock protein 27 in cytoskeletal remodeling of the airway smooth muscle cell, J. Appl. Physiol. 96, 1701–13.

Balland, M., Desprat, N., Icard, D., Fereol, S., Asnacios, A., Browaeys, J., Henon, S. and Gallet, F., 2006, Power laws in microrheology experiments on living cells: comparative analysis and modelling, Phys. Rev. E. 74, 021911.

Balland, M., Richert, A. and Gallet, F., 2005, The dissipative contribution of myosin II in the cytoskeleton dynamics of myoblasts, Eur. Biophys. J. 34, 255–61.

Bausch, A. R., Ziemann, F., Boulbitch, A. A., Jacobson, K. and Sackmann, E., 1998, Local measurements of viscoelastic parameters of adherent cell surfaces by magnetic bead microrheometry, Biophys. J. 75, 2038–49.

Bissig, H., Romer, S., Cipelletti, L., Trappea, V. and Schurtenbergera, P., 2003, Intermittent dynamics and hyper-aging in dense colloidal gels, Phys. Chem. Comm. 6, 21–3.

Bouchaud, J., 1992, Weak ergodicity breaking and aging in disordered systems, J. Phys. I. 2, 1705–13.

Brujic, J., Hermans, R., Walther, K. and Fernandez, J. M., 2006, Single-molecule force spectroscopy reveals signatures of glassy dynamics of the energy landscape of ubiquitin, Nature Physics. 2, 282–6.

Bursac, P., Fabry, B., Trepat, X., Lenormand, G., Butler, J. P., Wang, N., Fredberg, J. J. and An, S. S., 2007, Cytoskeleton dynamics: fluctuations within the network, Biochem. Biophys. Res. Commun. 355, 324–30.

Bursac, P., Lenormand, G., Fabry, B., Oliver, M., Weitz, D. A., Viasnoff, V., Butler, J. P. and Fredberg, J. J., 2005, Cytoskeletal remodelling and slow dynamics in the living cell, Nat. Mater. 4, 557–61.

Butler, J. P., Tolic-Norrelykke, I. M., Fabry, B. and Fredberg, J. J., 2002, Traction fields, moments, and strain energy that cells exert on their surroundings, Am. J. Physiol. Cell Physiol. 282, C595–605.

Byfield, F. J., Aranda-Espinoza, H., Romanenko, V. G., Rothblat, G. H. and Levitan, I., 2004, Cholesterol depletion increases membrane stiffness of aortic endothelial cells, Biophys. J. 87, 3336–43.

Chaudhuri, O., Parekh, S. H. and Fletcher, D. A., 2007, Reversible stress softening of actin networks, Nature 445, 295–8.

Cipelletti, L., Ramos, L., Manley, S., Pitard, E., Weitz, D. A., Pashkovski, E. E. and Johansson, M., 2003, Universal non-diffusive slow dynamics in aging soft matter, Faraday Discuss. 123, 237–51.

Cloitre, M., Borrega, R. and Leibler, L., 2000, Rheological aging and rejuvenation in microgel pastes, Phys. Rev. Lett. 85, 4819–22.

Crick, F. H. C. and Hughes, A. F. W., 1950, The physical properties of cytoplasm, Exp. Cell. Res. 1, 37–80.

Crocker, J. C., Valentine, M. T., Weeks, E. R., Gisler, T., Kaplan, P. D., Yodh, A. G. and Weitz, D. A., 2000, Two-point microrheology of inhomogeneous soft materials, Phys. Rev. Lett. 85, 888–91.

Dahl, K. N., Engler, A. J., Pajerowski, J. D. and Discher, D. E., 2005, Power-law rheology of isolated nuclei with deformation mapping of nuclear substructures, Biophys. J. 89, 2855–64.
Dembo, M. and Wang, Y. L., 1999, Stresses at the cell-to-substrate interface during locomotion of fibroblasts, Biophys. J. 76, 2307–16.
Deng, L., Trepat, X., Butler, J. P., Millet, E., Morgan, K. G., Weitz, D. A. and Fredberg, J., 2006, Fast and slow dynamics of the cytoskeleton, Nat. Mater. 5, 636–40.
Derec, C., Ducouret, G., Ajdari, A. and Lequeux, F., 2003, Aging and nonlinear rheology in suspensions of polyethylene oxide-protected silica particles, Phys. Rev. E 67, 061403.
Desprat, N., Richert, A., Simeon, J. and Asnacios, A., 2005, Creep function of a single living cell, Biophys. J. 88, 2224–33.
Dimitriadis, E. K., Horkay, F., Maresca, J., Kachar, B. and Chadwick, R. S., 2002, Determination of elastic moduli of thin layers of soft material using the atomic force microscope, Biophys. J. 82, 2798–810.
Discher, D. E., Janmey, P. and Wang, Y. L., 2005, Tissue cells feel and respond to the stiffness of their substrate, Science 310, 1139–43.
Engler, A. J., Sen, S., Sweeney, H. L. and Discher, D. E., 2006, Matrix elasticity directs stem cell lineage specification, Cell 126, 677–89.
Evans, E., 2001, Probing the relation between force – lifetime – and chemistry in single molecular bonds, Annu. Rev. Biophys. Biomol. Struct. 30, 105–28.
Fabry, B., Maksym, G. N., Butler, J. P., Glogauer, M., Navajas, D. and Fredberg, J. J., 2001a, Scaling the microrheology of living cells, Phys. Rev. Lett. 87, 148102.
Fabry, B., Maksym, G. N., Shore, S. A., Moore, P. E., Panettieri, R. A., Jr., Butler, J. P. and Fredberg, J. J., 2001b, Time course and heterogeneity of contractile responses in cultured human airway smooth muscle cells, J. Appl. Physiol. 91, 986–94.
Fabry, B., Maksym, G. N., Butler, J. P., Glogauer, M., Navajas, D., Taback, N. A., Millet, E. J. and Fredberg, J. J., 2003, Time scale and other invariants of integrative mechanical behavior in living cells, Phys. Rev. E. 68, 041914.
Findley, W. N., Lai, J. S. and Onaran, K., 1989, Creep and Relaxation of Nonlinear Viscoelastic Materials with an Introduction to Linear Viscoelasticity, Mineola, New York: Dover Publications, Inc.
Fredberg, J. J. and Stamenovic, D., 1989, On the imperfect elasticity of lung tissue, J. Appl. Physiol. 67, 2408–19.
Fuller, N. and Rand, R. P., 1999, Water in actin polymerization, Biophys. J. 76, 3261–6.
Gardel, M. L., Nakamura, F., Hartwig, J. H., Crocker, J. C., Stossel, T. P. and Weitz, D. A., 2006, Prestressed F-actin networks cross-linked by hinged filamins replicate mechanical properties of cells, Proc. Natl. Acad. Sci. U S A 103, 1762–7.
Gardel, M. L., Shin, J. H., MacKintosh, F. C., Mahadevan, L., Matsudaira, P. and Weitz, D. A., 2004, Elastic behavior of cross-linked and bundled actin networks, Science 304, 1301–5.
Gittes, F. and MacKintosh, F. C., 1998, Dynamic shear modulus of a semiflexible polymer network, Phys. Rev. E. 58, R1241–4.
Gunst, S. J. and Fredberg, J. J., 2003, The first three minutes: smooth muscle contraction, cytoskeletal events, and soft glasses, J. Appl. Physiol. 95, 413–25.
Gunst, S. J., Meiss, R. A., Wu, M. F. and Rowe, M., 1995, Mechanisms for the mechanical plasticity of tracheal smooth muscle, Am. J. Physiol. 268, C1267–76.
Henon, S., Lenormand, G., Richert, A. and Gallet, F., 1999, A new determination of the shear modulus of the human erythrocyte membrane using optical tweezers, Biophys. J. 76, 1145–51.
Hoffman, B. D., Massiera, G., Van Citters, K. M. and Crocker, J. C., 2006, The consensus mechanics of cultured mammalian cells, Proc. Natl. Acad. Sci. U S A 103, 10259–64.
Howard, J., 2001, Mechanics of Motor Proteins and the Cytoskeleton, Sunderland, Massachusetts: Sinauer Associates, Inc.
Ingber, D. E., 2006, Cellular mechanotransduction: putting all the pieces together again, Faseb. J. 20, 811–27.
Ingber, D. E., Dike, L., Hansen, L., Karp, S., Liley, H., Maniotis, A., McNamee, H., Mooney, D., Plopper, G., Sims, J. and et al., 1994, Cellular tensegrity: exploring how mechanical changes in

the cytoskeleton regulate cell growth, migration, and tissue pattern during morphogenesis, Int. Rev. Cytol. 150, 173–224.

Janmey, P. A. and Weitz, D. A., 2004, Dealing with mechanics: mechanisms of force transduction in cells, Trends Biochem. Sci. 29, 364–70.

Jones, W. R., Ting-Beall, P. H., Lee, G. M., Kelley, S. S., Hochmuth, R. M. and Guilak, F., 1999, Alterations in the Young's modulus and volumetric properties of chondrocytes isolated from normal and osteoarthritic human cartilage, J. Biomech. 32, 119–27.

Kabir, S. R., Yokoyama, K., Mihashi, K., Kodama, T. and Suzuki, M., 2003, Hyper-mobile water is induced around actin filaments, Biophys. J. 85, 3154–61.

Karcher, H., Lammerding, J., Huang, H., Lee, R. T., Kamm, R. D. and Kaazempur-Mofrad, M. R., 2003, A three-dimensional viscoelastic model for cell deformation with experimental verification, Biophys. J. 85, 3336–49.

Kasza, K. E., Rowat, A. C., Liu, J., Angelini, T. E., Brangwynne, C. P., Koenderink, G. H. and Weitz, D. A., 2007, The cell as a material, Curr. Opin. Cell Biol. 19, 101–7.

Kob, W., Donati, C., Plimpton, S. J., Poole, P. H. and Glotzer, S. C., 1997, Dynamical heterogeneities in a supercooled Lennard-Jones liquid, Phys. Rev. Lett. 79, 2827–30.

Lau, A. W. C., Hoffman, B. D., Davies, A., Crocker, J. C. and Lubensky, T. C., 2003, Microrheology, stress fluctuations, and active behavior of living cells, Phys. Rev. Lett. 91, 198101.

Laudadio, R. E., Millet, E. J., Fabry, B., An, S. S., Butler, J. P. and Fredberg, J. J., 2005, Rat airway smooth muscle cell during actin modulation: rheology and glassy dynamics, Am. J. Physiol. Cell. Physiol. 289, C1388–95.

Lenormand, G., Bursac, P., Butler, J. P. and Fredberg, J. J., 2007a, Out-of-equilibrium dynamics in the cytoskeleton of the living cell, Phys. Rev. E. 76, 041901.

Lenormand, G., Chopin, J., Bursac, P., Fredberg, J. J. and Butler, J. P., 2007b, Directional memory and caged dynamics in cytoskeletal remodelling, Biochem. Biophys. Res. Commun. 360, 797–801.

Lenormand, G., Millet, E., Fabry, B., Butler, J. P. and Fredberg, J. J., 2004, Linearity and time-scale invariance of the creep function in living cells, J. R. Soc. Interface 1, 91–7.

Levine, A. J. and Lubensky, T. C., 2001, Two-point microrheology and the electrostatic analogy, Phys. Rev. E 65, 11501.

Liu, A. J. and Nagel, S. R., 1998, Jamming is not just cool any more, Nature 396, 21–22.

Liu, J., Gardel, M. L., Kroy, K., Frey, E., Hoffman, B. D., Crocker, J. C., Bausch, A. R. and Weitz, D. A., 2006, Microrheology probes length scale dependent rheology, Phys. Rev. Lett. 96, 118104.

Mason, T. G., 2000, Estimating the viscoelastic moduli of complex fluids using the generalized Stokes-Einstein equation, Rheologica. Acta. 39, 371–8.

Mason, T. G. and Weitz, D. A., 1995a, Linear Viscoelasticity of Colloidal Hard Sphere Suspensions near the Glass Transition, Phys. Rev. Lett. 75, 2770–3.

Mason, T. G. and Weitz, D. A., 1995b, Optical measurements of frequency-dependent linear viscoelastic moduli of complex fluids, Phys. Rev. Lett. 74, 1250–3.

Matthews, B. D., Overby, D. R., Alenghat, F. J., Karavitis, J., Numaguchi, Y., Allen, P. G. and Ingber, D. E., 2004, Mechanical properties of individual focal adhesions probed with a magnetic microneedle, Biochem. Biophys. Res. Commun. 313, 758–64.

Mijailovich, S. M., Kojic, M., Zivkovic, M., Fabry, B. and Fredberg, J. J., 2002, A finite element model of cell deformation during magnetic bead twisting, J. Appl. Physiol. 93, 1429–1436.

Mizuno, D., Tardin, C., Schmidt, C. F. and Mackintosh, F. C., 2007, Nonequilibrium mechanics of active cytoskeletal networks, Science 315, 370–3.

Nash, G. B. and Gratzer, W. B., 1993, Structural determinants of the rigidity of the red cell membrane, Biorheology 30, 397–407.

Ohayon, J., Tracqui, P., Fodil, R., Fereol, S., Laurent, V. M., Planus, E. and Isabey, D., 2004, Analysis of nonlinear responses of adherent epithelial cells probed by magnetic bead twisting: A finite element model based on a homogenization approach, J. Biomech. Eng. 126, 685–98.

Overby, D. R., Matthews, B. D., Alsberg, E. and Ingber, D. E., 2005, Novel dynamic rheological behavior of individual focal adhesions measured within single cells using electromagnetic pulling cytometry, Acta. Biomater. 1, 295–303.
Pratusevich, V. R., Seow, C. Y. and Ford, L. E., 1995, Plasticity in canine airway smooth muscle, J. Gen. Physiol. 105, 73–94.
Puig-de-Morales, M., Millet, E., Fabry, B., Navajas, D., Wang, N., Butler, J. P. and Fredberg, J. J., 2004, Cytoskeletal mechanics in adherent human airway smooth muscle cells: probe specificity and scaling of protein-protein dynamics, Am. J. Physiol. Cell. Physiol. 287, C643–54.
Puig-de-Morales-Marinkovic, M., Turner, K. T., Butler, J. P., Fredberg, J. J. and Suresh, S., 2007, Viscoelasticity of the human red blood cell, Am. J. Physiol. Cell Physiol. 293, C597–605.
Ramos, L. and Cipelletti, L., 2001, Ultraslow dynamics and stress relaxation in the aging of a soft glassy system, Phys. Rev. Lett. 87, 245503.
Rico, F., Roca-Cusachs, P., Gavara, N., Farre, R., Rotger, M. and Navajas, D., 2005, Probing mechanical properties of living cells by atomic force microscopy with blunted pyramidal cantilever tips, Phys. Rev. E Stat. Nonlin. Soft Matter Phys. 72, 021914.
Roca-Cusachs, P., Almendros, I., Sunyer, R., Gavara, N., Farre, R. and Navajas, D., 2006, Rheology of passive and adhesion-activated neutrophils probed by atomic force microscopy, Biophys. J. 91, 3508–18.
Rowat, A. C., Lammerding, J. and Ipsen, J. H., 2006, Mechanical properties of the cell nucleus and the effect of emerin deficiency, Biophys. J. 91, 4649–64.
Sato, M., Theret, D. P., Wheeler, L. T., Ohshima, N. and Nerem, R. M., 1990, Application of the micropipette technique to the measurement of cultured porcine aortic endothelial cell viscoelastic properties, J. Biomech. Eng. 112, 263–8.
Schmid-Schonbein, G. W., Sung, K. L., Tozeren, H., Skalak, R. and Chien, S., 1981, Passive mechanical properties of human leukocytes, Biophys. J. 36, 243–256.
Smith, B. A., Roy, H., De Koninck, P., Grutter, P. and De Koninck, Y., 2007, Dendritic spine viscoelasticity and soft-glassy nature: balancing dynamic remodeling with structural stability, Biophys. J. 92, 1419–30.
Smith, B. A., Tolloczko, B., Martin, J. G. and Grutter, P., 2005, Probing the viscoelastic behavior of cultured airway smooth muscle cells with atomic force microscopy: stiffening induced by contractile agonist, Biophys. J. 88, 2994–3007.
Smith, P. G., Deng, L., Fredberg, J. J. and Maksym, G. N., 2003, Mechanical strain increases cell stiffness through cytoskeletal filament reorganization, Am. J. Physiol. Lung. Cell Mol. Physiol. 285, L456–63.
Sollich, P., 1998, Rheological constitutive equation for a model of soft glassy materials, Phys. Rev. E. 58, 738–59.
Sollich, P., Lequeux, F., Hébraud, P. and Cates, M. E., 1997, Rheology of soft glassy materials, Phys. Rev. Lett. 78, 2020–3.
Stamenovic, D., 2006, Cell mechanics: two regimes, maybe three? Nat. Mater 5, 597–8.
Stamenovic, D. and Coughlin, M. F., 1999, The role of prestress and architecture of the cytoskeleton and deformability of cytoskeletal filaments in mechanics of adherent cells: a quantitative analysis, J. Theor. Biol. 201, 63–74.
Stamenovic, D., Rosenblatt, N., Montoya-Zavala, M., Matthews, B. D., Hu, S., Suki, B., Wang, N. and Ingber, D. E., 2007, Rheological behavior of living cells is timescale-dependent, Biophys. J. 93, L39–41.
Stamenovic, D., Suki, B., Fabry, B., Wang, N. and Fredberg, J. J., 2004, Rheology of airway smooth muscle cells is associated with cytoskeletal contractile stress, J. Appl. Physiol. 96, 1600–5.
Struick, L. C. E., 1978, Physical Aging in Amorphous Polymers and Other Materials, Houston, Texas: Elsevier.
Suzuki, M., Kabir, S. R., Siddique, M. S., Nazia, U. S., Miyazaki, T. and Kodama, T., 2004, Myosin-induced volume increase of the hyper-mobile water surrounding actin filaments, Biochem. Biophys. Res. Commun. 322, 340–6.

Theret, D. P., Levesque, M. J., Sato, M., Nerem, R. M. and Wheeler, L. T., 1988, The application of a homogeneous half-space model in the analysis of endothelial cell micropipette measurements, J. Biomech. Eng. 110, 190–9.

Thoumine, O., Cardoso, O. and Meister, J. J., 1999, Changes in the mechanical properties of fibroblasts during spreading: a micromanipulation study, Eur. Biophys. J. 28, 222–34.

Thoumine, O. and Ott, A., 1997, Time scale dependent viscoelastic and contractile regimes in fibroblasts probed by microplate manipulation, J. Cell Sci. 110 (Pt 17), 2109–16.

Trappe, V., Prasad, V., Cipelletti, L., Segre, P. N. and Weitz, D. A., 2001, Jamming phase diagram for attractive particles, Nature 411, 772–5.

Trepat, X., Deng, L., An, S. S., Navajas, D., Tschumperlin, D. J., Gerthoffer, W. T., Butler, J. P. and Fredberg, J. J., 2007, Universal physical responses to stretch in the living cell, Nature 447, 592–5.

Tseng, Y., Kole, T. P. and Wirtz, D., 2002, Micromechanical mapping of live cells by multiple-particle-tracking microrheology, Biophys. J. 83, 3162–76.

Tseng, Y. and Wirtz, D., 2001, Mechanics and multiple-particle tracking microheterogeneity of alpha-actinin-cross-linked actin filament networks, Biophys. J. 81, 1643–56.

Valentine, M. T., Perlman, Z. E., Mitchison, T. J. and Weitz, D. A., 2005, Mechanical properties of Xenopus egg cytoplasmic extracts, Biophys. J. 88, 680–9.

Van Citters, K. M., Hoffman, B. D., Massiera, G. and Crocker, J. C., 2006, The role of F-actin and myosin in epithelial cell rheology, Biophys J. 91, 3946–56.

Wang, N., Butler, J. P. and Ingber, D. E., 1993, Mechanotransduction across the cell surface and through the cytoskeleton, Science 260, 1124–7.

Wang, N., Naruse, K., Stamenovic, D., Fredberg, J. J., Mijailovich, S. M., Tolic-Norrelykke, I. M., Polte, T., Mannix, R. and Ingber, D. E., 2001, Mechanical behavior in living cells consistent with the tensegrity model, Proc. Natl. Acad. Sci. U S A 98, 7765–70.

Wang, N., Tolic-Norrelykke, I. M., Chen, J., Mijailovich, S. M., Butler, J. P., Fredberg, J. J. and Stamenovic, D., 2002, Cell prestress. I. Stiffness and prestress are closely associated in adherent contractile cells, Am. J. Physiol. Cell Physiol. 282, C606–16.

Weeks, E. R., Crocker, J. C., Levitt, A. C., Schofield, A. and Weitz, D. A., 2000, Three-dimensional direct imaging of structural relaxation near the colloidal glass transition, Science 287, 627–31.

Wottawah, F., Schinkinger, S., Lincoln, B., Ananthakrishnan, R., Romeyke, M., Guck, J. and Kas, J., 2005, Optical rheology of biological cells, Phys. Rev. Lett. 94, 098103.

Yamada, S., Wirtz, D. and Kuo, S. C., 2000, Mechanics of living cells measured by laser tracking microrheology, Biophys. J. 78, 1736–47.

Yanai, M., Butler, J. P., Suzuki, T., Sasaki, H. and Higuchi, H., 2004, Regional rheological differences in locomoting neutrophils, Am. J. Physiol. Cell. Physiol. 287, C603–11.

Unexpected Linkage Between Unstirred Layers, Exclusion Zones, and Water

Gerald H. Pollack and James Clegg

Abstract We make the case that the unstirred layers of classical physiology arise from the influence of surfaces on the structure, and therefore the properties, of contiguous water. Traditionally, unstirred layers have been thought to arise merely out of stagnant volumes adjacent to membranes and other surfaces. These volumes would have to extend tens to hundreds of micrometers in order to account for the observed effects. On the other hand, charged and hydrophilic surfaces have been shown to impact water out to surprising distances, on the order of hundreds of micrometers. We present evidence that it is this water-structuring effect that is responsible for the unstirred layer.

Keywords Unstirred layers · exclusion zones · water structure

1 Unstirred Layers (USLs)

It is generally agreed that unstirred layers (USLs) are more or less static fluid regions adjacent to membranes, and probably many other surfaces, within which thermal convection or density gradients do not cause significant mixing (Barry and Diamond, 1984). Even when the surrounding fluid is vigorously stirred this seemingly static fluid remains, at least to some extent, and there is a relationship between the thickness of USLs (δ) and the intensity of stirring of the surrounding fluid (Barry and Diamond, 1984).

USLs are of critical importance in a wide variety of physiological processes, notably those involving solute transport, enzyme and other chemical reactions on or near the cell surface. Membrane electrochemical potentials may also be affected because concentrations of ions and other solutes adjacent to the plasma membrane are not the same as those in the bulk, an important point to which we will return. The literature on USLs is massive, including research in physical chemistry where the USLs have been referred to as boundary layers (House, 1974; Barry and Diamond, 1984). These

G.H. Pollack
Department of Bioengineering, Box 355061, University of Washington, Seattle WA, 98195, USA
e-mail: ghp@u.washington.edu

G.H. Pollack, W.-C. Chin (eds.), *Phase Transitions in Cell Biology*,

layers have been implicated in many processes, and in a wide variety of systems such as the gills of mussels (Wright et al., 1980), the human jejunum (Pappenheimer, 2001), monolayer cultures of mammalian cells (Spivak et al., 2006), rainbow-trout embryos and larvae (Ciuhandu et al., 2007) and a giant-celled alga (Kim et al., 2006). We make no pretence about a thorough review of this literature but will attempt to point out features that are salient to the topics covered in this essay.

A brief examination of the history of USLs seems worthwhile. In an important early review, Dainty and House (1966) state that the USL concept was originally developed by Noyes and Whitney in 1897 and further in 1904 by Nernst. However, based on the translation of Wilhelm Pfeffer's 1877 book, *Osmotische Unteruchungen* (Kepner and Stadelmann, 1985), it appears that Pfeffer described USLs well before those authors. Pfeffer clearly emphasized the critical role played by USLs in his extensive research on osmosis and related processes. Key papers on USLs in the early to mid-1900s have been considered or reviewed by Teorell (1936) Ginzberg and Katchalsky (1963) and Dick (1959). The review by Barry and Diamond, (1984) is definitive in our opinion, with excellent overall coverage of USLs. More recently, work from the laboratory of Peter Pohl considers the wide-ranging significance and implications of USLs and their relation to what is called reaction layers, RLs (see Antonenko et al., 1996; Pohl et al., 1998). RLs are solution layers adjacent to cell membranes within which chemical reactions, including enzymatic ones, take place. These layers have been described previously (see Gutnecht and Tosteson,1973; Verkman and Dix, 1984) but particularly by Antonenko et al. (1996). We mention them here to draw attention to the important relationship between USLs and RLs and, as we will propose, the importance of water in the operation of both layers.

A key feature of USLs concerns their thickness—the distance from membranes and other surfaces over which mechanical stirring is not effective. An excellent source of information on USL thickness (δ) is the review by Barry and Diamond (1984). Without stirring, δ exceeds 300 μm, depending on the nature of the interface, and can even extend over several millimeters from the surface. Despite the most vigorous stirring, δ remains near 10 μm, obviously a huge distance in molecular terms. The view is widely held that the USL is not stationary but a region of slow laminar flow, parallel to the interface, in which solute transport occurs by diffusion alone (Dainty and House, 1966).

A key question arises: what generates the USL? A logical possibility is that the physical properties of the USL are somehow different from those of regions beyond. To our knowledge the "structure" of water in such interfacial regions has not been considered in this connection. We wish to raise this possibility, that the influence of the adjacent "solid" surface, membrane, or otherwise, alters the structure and properties of the interfacial water in such a way that the USL is generated.

2 Cellular Water

Studies on the properties and structure of water in cells and adjacent to surfaces have been many, and the topic has had a long and often controversial history. It is not appropriate here to review that history, except to note certain points. Cell biology seems to have proceeded throughout its history on the assumption that all but a small fraction (perhaps 5%) of total cell water exhibited the properties of bulk water, and that the cytoplasm could be treated as an ordinary, although non-ideal, aqueous solution (see Clegg, 1982). Some have objected to that view over the years, notably Gilbert Ling, Walter Drost-Hansen, and Aphanasij Troschin, who proposed that most and probably all of cell water is different. Those three scientists began their work in the 1950s, and two continue their work today (see Pollack et al., 2006). A high point arrived in the late 1960s when both Freeman Cope (1969), and Carlton Hazlewood and colleagues (1969), published papers using pulsed NMR technology. They concluded that the structure of most or even all of water in skeletal muscle and brain was not that of the bulk liquid.

Much attention was given to these results, supporting the positions of Ling, Drost-Hansen and Troschin, but it was not very long until opposition to those views arose (see Kolata, 1976). What followed was a contentious period spanning some 10 years of claim and counterclaim (considered to some extent in works by Hazlewood, 1973; Franks and Mathias, 1982; Clegg and Drost-Hansen, 1991; Drost-Hansen and Clegg, 1979; Ling, 2001; Mentré, 2001; Pollack, 2001; Pollack et al., 2006). As of this writing research on cell water has slowed greatly, with a few notable exceptions (see articles in Mentré, 2001; Pollack, 2001; Pollack et al., 2006) although interest in the subject continues (for example, Leterrier, 2001; Shepherd, 2006). The version given in textbooks and the general literature is not much different than it was a century ago: almost all of cell water is considered to be essentially the same as that in the bulk liquid. But there remains a dissenting group, albeit small, that believes the last word has yet to be said; we are among that group.

That surfaces alter the structure of water immediately adjacent to them is well established for very short distances, equivalent to two or three molecular layers of water. But it is our impression that very few (other than those named above) would accept the possibility that such influences might extend over distances of tens or hundreds of nanometers, let alone micrometers. However, in the sections that follow we will make a case for just that; namely, that the existence and properties of USLs arise from the influence of membranes and other "solid" surfaces on the structure of water adjacent to them.

3 Exclusion Zones (EZs)

Motivated independently from unstirred layer considerations, studies of the physical behavior of this near-surface aqueous zone were begun several years ago (Zheng and Pollack, 2003). These studies were in fact broader than just examining the unstirred layer next to cells: they dealt with the nature of aqueous regions adjacent to many hydrophilic surfaces.

A half-century old review of the solid-liquid interface (Henniker, 1949) cited more than 100 studies showing a surface influence extending into the liquid far beyond the two or three molecular layers generally implied; and, a later study showed that on polished quartz, ordered water could extend out to some 600 molecular layers from the surface (Pashley and Kitchener, 1979). For many decades Drost-Hansen has mustered evidence that similar distances are involved in a wide variety of surface systems, the adjacent water being referred to as vicinal water (see Drost-Hansen, 2006). These reports, and many references therein, indicate considerably more structuring than has generally been accepted. Hence, positions were taken on both sides of the issue.

The experimental approach taken by Zheng and Pollack (2003) was to explore the behavior of colloidal particles next to hydrophilic surfaces. Since it is energetically favorable for colloidal particles and solutes to reside outside of structured regions so they can more readily hydrate, attempts were made to examine solute partitioning in the vicinity of interfaces. If such solutes are preferentially excluded from the interfacial zone, the presumption was that this zone was likely to be structured differently from ordinary bulk water.

Initial results seemed astonishing (Zheng and Pollack, 2003). A polyvinyl alcohol gel was placed in a chamber, and the edge of the gel was visualized in an optical microscope. A suspension of latex microspheres was poured over the gel, filling the chamber. Immediately, microspheres began migrating away from the gel surface, into the water. Within several minutes a microsphere-free zone could be seen extending some 50–100 μm from the gel surface (Fig. 1). The result implied that the exclusion zone could quite possibly extend by as much as 100 μm from the gel surface.

In retrospect, this observation should not have been quite so astonishing. As mentioned, Drost-Hansen has long argued that the properties of water adjacent to surfaces, vicinal water, were different from the bulk, and that these properties extended over great distances from surfaces, in some cases micrometers (see Drost-Hansen,

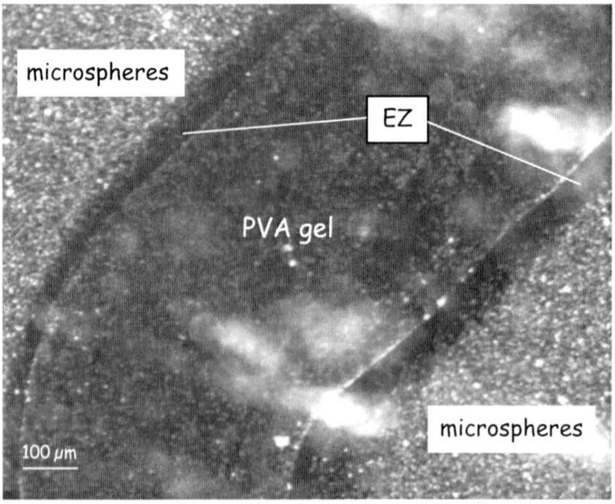

Fig. 1 Exclusion zones (*dark*, "EZ") on either side of a polyvinyl alcohol gel. Microspheres (*upper left* and *lower right*) are excluded from the region close to the gel. Cloud-like bright areas are artifactual reflections

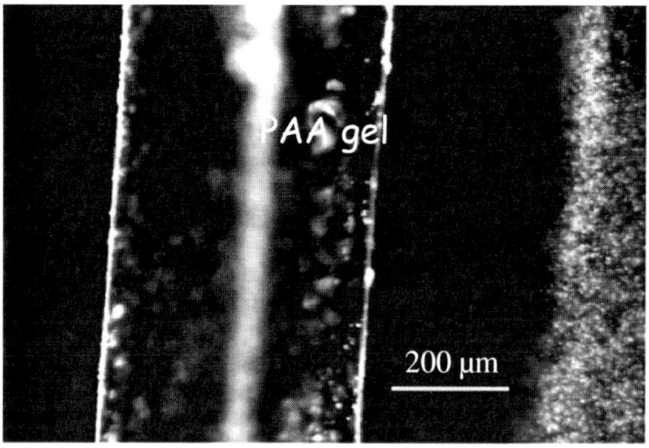

Fig. 2 Exclusion zones (*dark*) on either side of a polyacrylic acid gel. Microspheres translated from the gel surface appear at the far right. The vertical bright stripe in the middle of the PAA gel is an optical artifact

2006). Further, those observations seemed in concordance with the many results obtained during the first half of the previous century, as reviewed by Henniker (1949). But the fact remains that these observations and proposals have had little effect on current views concerning interfacial water.

Even more remarkable than the 100-μm EZ seen with polyvinyl alcohol gels was the exclusion zone seen with the more highly charged polyacrylic-acid gels. There, the microsphere-exclusion zone could easily extend from the gel surface by 250 μm or more (Fig. 2).

It was not only microspheres that were excluded from the region next to gels. Many suspended or dissolved species were excluded, ranging in size from particles such as sand and silt, to biological entities including bacteria and erythrocytes, down to much smaller species such as proteins and low molecular weight dyes. All of those were excluded (Zheng et al., 2006). The kinds of surfaces that produced such exclusion also ranged broadly. They included a series of hydrogels, various polymeric surfaces including Nafion, and biological surfaces including vascular endothelium, seaweed, and various plant roots. Muscle, for example, produced an EZ of several hundred micrometers (Zheng et al., 2006). Hence, the exclusion phenomenon seems rather general, and is not the quirk of a particular solute juxtaposed next to a particular surface.

It was after those experiments had been largely completed that we came upon the study by Green and Otori (1970) which linked surfaces and microspheres with unstirred layers. Curious about the nature of the unstirred layer around the eye cornea, those investigators exposed the cornea to suspensions of polystyrene latex microspheres in much the same manner as we had done more recently with other hydrophilic surfaces. The results were similar. Green and Otori found microsphere-exclusion zones on the order of 350 μm. They found exclusion not only adjacent to the natural cornea, but also next to artificial contact lenses – results virtually identical to what we found next to polyNIPAM, the gel used to fabricate contact lenses. Clearly, the results of Green and Otori and the more recent results obtained in the Pollack laboratory are for all intents and purposes indistinguishable. The unstirred layer of classical physiology and the solute-exclusion zone seem to be one and the same.

4 Nature of Unstirred Layers and Exclusion Zones

What is the nature of the water within USLs and EZs? One possibility is that it is identical to bulk water, and that exclusion merely derives from proximity to surfaces. Because of entropic effects, solutes should be in lower concentrations in regions "close to" surfaces. Such effects are generally considered to extend out to several, or even tens of molecular layers of water. But, here the distances under consideration are far greater: a span of 250–300 μm is equivalent to a stack of about a million water molecules. Such numbers are far too large to fall into the realm of surface-entropic effects.

A second possibility is that the water in this zone may be impacted by the surface; i.e., that the surface influences water structure out to distances considerably farther than expected. The results described above are fully consistent with that possibility, but they do not prove it. More recent results, on the other hand, lend further support to the possibility that water in the exclusion zone is physically different from bulk water.

A significant indication of water-structural differences in the vicinal zone derives from studies using UV-Vis spectrometry (Zheng et al., 2006). Those studies involved Nafion, whose surface generates an appreciable exclusion zone. A sheet of Nafion was glued to the inside face of one of the four vertical walls of a standard cuvette, which was then filled with distilled, deionized water. The spectrometer was modified such that a narrow slit of light could sample zones parallel to, and at variable distances from, the Nafion face.

Far from the Nafion surface, the absorption spectrum was indistinguishable from that of bulk water. As the narrow sampling window came progressively closer to the Nafion surface, a new absorption peak at ∼270 nm, could be detected. The peak became prominently visible at the exclusion-zone boundary, ∼ 500 μm from the Nafion surface. It grew further as the window came closer to the surface, ultimately dominating the pattern (Fig. 3). Since no absorption peaks at 270 nm are found in bulk water, the presence of a strong "anomalous" absorption peak implies that exclusion-zone water differs appreciably from bulk water.

Artifacts in such measurements are possible: namely, some leaching of polymer from the Nafion could be responsible for this absorption. Control experiments have been carried out to determine whether a similar absorption peak occurs in situations in which water structuring is anticipated. Indeed, various concentrated salt solutions and sugar solutions showed similar absorption peaks (Chai et al., 2008). When the salts were first heat-purified in order to remove volatile hydrocarbon contaminants,

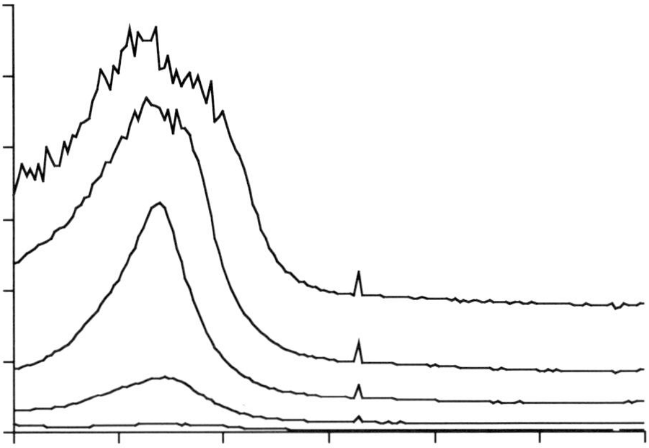

Fig. 3 UV-Vis absorption spectrum as a function of distance from the Nafion surface

the absorption peaks grew more intense. Hence, these absorption peaks do not appear to be artifacts, but may be a genuine "signature" of water structure.

Several other physical features of exclusion-zone water differ from their counterparts in bulk water. These include shorter NMR-T_2 relaxation times, reduced infrared radiation, and a steep electrical potential gradient (Zheng et al., 2006). Unpublished experiments also show different optical refractive features, increased viscosity, and higher density.

Taken together, these observations imply that water in the exclusion zone is physically, and perhaps chemically different from ordinary bulk water. In particular, the reduced infrared radiation and shorter T_2 values imply that water in USLs and EZs may be structurally more stable than bulk water—perhaps partially liquid crystalline in nature. In such a case it would be clear why colloids and solutes might be excluded: energetically, they prefer to lodge in the region of bulk water, where they can more easily hydrate. It also accounts for the cornea and contact lens observations of Green and Otori (1970), who found that microspheres did not invade the unstirred layer. Apparently, the microspheres avoided the region because it was structured in a way that did not admit them.

Thus, the unstirred layers of classical physiology appear to be the same as exclusion zones found next to many hydrophilic surfaces. Water in this zone is extensively impacted by the surface, and the resulting structure tends to exclude particles and solutes. On the other hand, important questions remain. One of the issues to be studied in more detail is the degree to which ions and small solutes penetrate the exclusion zone. To our knowledge this has not yet been explored directly. Although larger solutes and colloids seem fully excluded from such zones, it is not yet clear whether the smallest solutes are fully excluded, partially excluded, or not excluded. Given the striking similarities thus far established between unstirred layers and exclusion zones, there is reason to be optimistic that the "limited diffusion" of small solutes through the USL may be seen as well in the exclusion zone, but such hypotheses remain to be tested.

5 Ultrastructural Evidence for Exclusion Zones

Mollenhauer and Morré (1978) seem to have been the first to point out that zones of exclusion (their term) can be detected at the ultrastructural level in eukaryotic cells using transmission electron microscopy. They defined these zones as "differentiated regions of cytoplasm in which ribosomes, glycogen and organelles such as mitochondria, plastids or microbodies are scarce or absent." They also list other cell structures that are excluded. Our impression is that exclusion zones are consistently found adjacent to various membranes or cytoskeletal structures. These include actin filaments (Kamitsubo, 1972), and microtubules (Stebbings and Willison, 1973; Stebbings and Hunt, 1982), and they persist even when most of the water has been osmotically removed (Albrecht-Buehler and Bushnell, 1982). Thus, we believe these ultrastructural exclusion zones are comparable to the USLs and EZs we have discussed in this paper, extending their importance to the *in vivo* case.

6 Concluding Comments

The USL of classical physiology has long been known and studied. That trail has gone on with little or no attention given to what generates it. In this essay we support the view that the properties and structure of both the USL and EZ arise because of the influence of surfaces on the properties and structure of water in its vicinity. This water is somehow altered. The alteration may be profound—enough so that it has been considered to constitute a "fourth phase" of water (Hardy, 1932). The view that "boundary water" is altered is supported by evidence briefly summarized here, much of it coming from work done over the last 50 years, indicating that the influence of surfaces extends over large distances, exceeding 100 μm. That contention, standing in opposition to the conventional wisdom, indicates that the role of water in and around cells and tissues is vastly more important than the teachings of current textbooks permit. As always, the evidence will prevail, and the future will decide that outcome.

References

Albrecht-Buehler G, Bushnell A (1982) Reversible compression of the cytoplasm. Exp Cell Res 140:173–189

Antonenko YN, Pohl P, Rosenfeld E (1996) Visualization of the reaction layer in the immediate membrane vicinity. Arch Biochem Biophys 333:225–232

Barry PH, Diamond JM (1984) Effects of unstirred layers on membrane phenomena. Physiol Rev 64:763–871

Chai, B-H, Zheng, J-M, Zhao, Q, Pollack, GH (2008) Spectroscopic studies of solutes in aqueous solution. J. Phys. Chem A 112:2242–2247

Ciuhandu CS, Wright PA, Goldberg JI, Stevens ED (2007) Parameters influencing the dissolved oxygen in the boundary layer of rainbow trout (*Oncorhynchus mykiss*) embryos and larvae. J Exp Biol 210:1435–1445

Clegg JS (1982) Alternative views on the role of water in cell function. In: Franks F, Mathias SF (ed.) *Biophysics of Water*. John Wiley and Sons, New York, pp. 365–385
Clegg JS, Drost-Hansen W (1991) On the biochemistry and cell physiology of water. In: Hochachka PW, Mommsen TP (ed.), *Biochemistry and Molecular Biology of Fishes*, Elsevier, Amsterdam, pp. 1–23
Cope F (1969) Nuclear magnetic resonance evidence using D_2O for structured water in muscle and brain. Biophys J 9:303–319
Dainty J, House CR (1966) "Unstirred layers" in frog skin. J Physiol 182:66–78
Dick DAT (1959) Osmotic properties of living cells. Int Rev Cytol 8:387–448
Drost-Hansen W (2006) Vicinal hydration of biopolymers: cell biological consequences. In: Pollack GH, Cameron IL, Wheatley DN (ed.). *Water and the Cell*. Springer-Verlag, Berlin, pp. 175–217
Drost-Hansen W, Clegg JS (ed.) (1979) *Cell Associated Water*. Ascademic Press, New York, p. 440
Franks F, Mathias SF (ed.) (1982) *Biophysics of Water*. John Wiley and Sons, New York, p. 400
Ginzberg BZ, Katchalsky A (1963) The frictional coefficients of the flows of non-electrolytes through artificial membranes. J Gen Physiol 47:403–418
Green K, Otori T (1970) Direct measurement of membrane unstirred layers. J Physiol 207:93–102
Gutknecht, J, Tosteson, DC (1973). Diffusion of weak acids across lipid bilayer membranes: effects of chemical reactions in the unstirred layers. Science 182:1258–1261
Hardy, W (1932) Problems of the boundary state. Phil Trans Roy Soc Lon Ser A 230:1–37
Hazlewood CF (ed.) (1973) Physicochemical state of ions and water in living tissues and model systems. Ann NY Acad Sci 204:5–631
Hazlewood CF, Nichols BL, Chamberlain NF (1969) Evidence for the existence of a minimum of two phases of ordered water in skeletal muscle. Nature 222:747–750
Henniker, JC (1949) The depth of the surface zone of a liquid. Rev Mod Phys 21(2):322–341
House CR (1974) Water Transport in Cells. Arnold Press, London, p. 276
Kamitsubo, E (1972) Motile protoplasmic fibrils in cells of the *Characae*. Protoplasma 74:53–70
Kepner GR, Stadelmann EJ (1985) *Translation of Osmotische Untersuchungen. Studien zur Zellemechanik*. Van Nostrand Reinhold, New York, p. 267
Kim Y, Ye Q, Reinhardt H, Steudle E (2006) Further quantification of the role of internal stirred layers during the measurement of transport coefficients in giant internodes of *Chara* by a new stop-flow technique. J Exp Bot 57:4133–4144
Kolata G (1976) Water structure and ion binding: a role in cell physiology? Science 192:1220–1222
Leterrier JF (2001) Water and the cytoskeleton. Cell Mol Biol (Noisy-le-grand) 47:901–923
Ling GN (2001) Life at the Cell and Below-Cell Level. Pacific Press, New York, p. 373
Mentré P (ed.) (2001) Water in the cell. Cell Mol Biol 47:709–970
Mollenhauer HH, Morré DJ (1978) Structural compartmentation of the cytosol: zones of exclusion, zones of adhesion, cytoskeletal and intercisternal elements. In: Roodyn DB (ed.) *Subcellular Biochemistry*, vol. 5, Plenum Press, New York, pp. 327–362
Pappenheimer JR (2001) Role of pre-epithelial "unstirred" layers in absorption of nutrients from the human jejunum. J Membrane Biol 179:185-204
Pashley RM, Kitchener JA (1979) Surface forces in adsorbed multilayers of water on quartz. J Coll Interface Sci 71: 491–500
Pohl P, Saparov SM, Antonenko YN (1998) The size of the unstirred layer as a function of the solute diffusion coefficient. Biophys J 75:1403–1409
Pollack GH (2001) *Cells, Gels and the Engines of Life*. Ebner and Sons, Seattle, p. 305
Pollack GH, Cameron IL, Wheatley DN (ed.) (2006) *Water and the Cell*. Springer-Verlag, Berlin, pp. 165–174.
Shepherd VA (2006) The cytomatrix as a cooperative system of macromolecular and water networks. Curr Top Dev Biol 75:171–233
Spivak CE, Oz M, Beglan CL, Shrager RI (2006) Diffusion delays and unstirred layer effects at monolayer cultures of Chinese hamster ovary cells. Cell Biochem Biophys 45:43–58
Stebbings H and Hunt C (1982) The nature of the clear zone around microtubules. Cell and Tissue Res 227:609–617

Stebbings H, Willison JHM (1973) Structure of microtubules: a study of freeze-etched and negatively stained microtubules from the ovaries of *Notonecta*. Z. Zellforsch u Mikrosc Anat 138(3):387–396

Teorell T (1936) A method of studying conditions within diffusion layers. J Biol Chem 113:735–748

Verkman AS, Dix JA (1984) Effect of unstirred layers on binding and reaction kinetocs at a membrane surface. Anal Biochem 142:109–116

Wright SH, Becker SA, Stephens GC (1980) Influence of temperature and unstirred layers on the kinetics of glycine transport in isolated gills of *Mytilus californianus*. J Exp Zool 214:27–35

Zheng, JM, Pollack, GH (2003) Long range forces extending from polymer surfaces. Phys Rev E.: 68:031408

Zheng, J-M, Chin, W-C, Khijniak, E, Khijniak, E, Jr, Pollack, GH (2006) Surfaces and Interfacial Water: Evidence that hydrophilic surfaces have long-range impact. Adv. Colloid Interface Sci. 127:19–27

"Autothixotropy" of Water – An Unknown Physical Phenomenon, and its Possible Importance for the Cytoskeleton

Bohumil Vybíral and Pavel Voráček

Abstract A series of until now unknown phenomena ongoing in water was discovered in laboratory experiments, which made gravimetric measurements, with the necessarily extreme precision, impossible. This behaviour of water, which we call *autothixotropy*, is the subject of the presented report on the experimental research. We also propose a possible explanation. Finally, the dependence of the manifestation of autothixotropy on the degree of molecular translative freedom is mentioned and its possible relevance to the cytoskeleton is pointed out.

Keywords Water · quasi-staticfriction · autothixotropy · cytoskeleton · ion-kernels

1 Laboratory Observations

From 1978 to 1986 a series of measurements were performed (Vybíral, 1987, 1989) in order to verify the gravitational law in fluids as deduced by Horák (1984). In 1978 a peculiar phenomenon was observed in the measurements which compelled the author to use another method.

In a private communication between the authors, the new – then disturbing – phenomenon was discussed and a series of experiments focusing on this phenomenon were proposed. In the Department of Physics of Pedagogical Faculty of University of Hradec Králové, the original experimental device was reconstructed with the purpose of creating the best conditions for the observation of the phenomenon (Fig. 1).

After objects immersed in the water have been at rest for one or several days, seven qualitatively different phenomena can be observed:

B. Vybíral
Department of Physics, Pedagogical Faculty, University of Hradec Králové, Rokitanského 62, CZ-500 03 Hradec Králové, Czech Republic
e-mail: Bohumil.Vybiral@uhk.cz

This article appeared, in its preliminary form, in July 2003 on the web at the address: http://arxiv.org/abs/physics/0307046

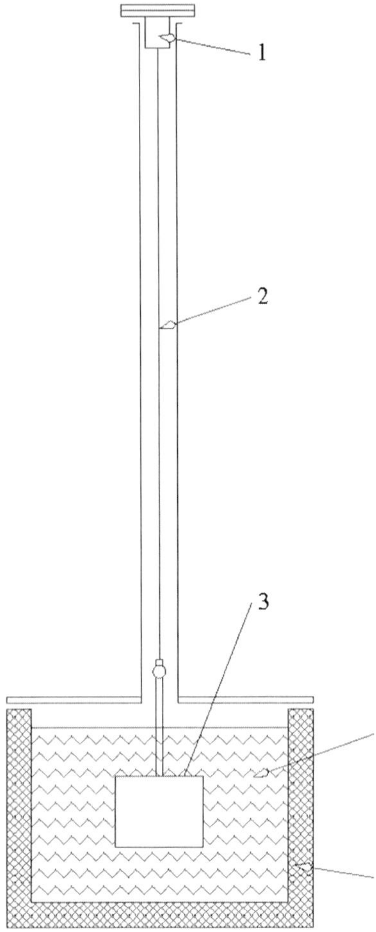

Fig. 1 Schematic drawing of the experimental device: 1 – fine turning hanger; 2 – phosphor-bronze filament (cross-section 0.20 mm × 0.025 mm, length 0.811 m); 3 – stainless steel plate (b. 60 mm, h. 50 mm, t. 1 mm); 4 – distilled air-free water; 5 – heat-insulated closed sterilized vessel (inner diameter 150 mm)

(1) When the hanger is rotated by a certain angle, the plate immersed in the water remains in practically the same position, in spite of the twisting tension arising in the thin filament. When a certain critical angle is reached, the plate will rotate – relatively quickly – into a new neutral position determined by the hanger, which means into the position where the filament is relaxed (i.e. with no torsion). If the rotation of the hanger is interrupted before the critical angle is reached, a "creep" toward a new neutral position can be observed after some days or weeks. When a smooth-surfaced cylinder, capable of rotating around its own axis, was used instead of the plate, these phenomena were not observed.

(2) Another phenomenon, though – in the clarity of appearance – weaker than the phenomenon mentioned above, can simultaneously be observed: an immediate

rotation of the plate in the direction of the rotation of the hanger; nevertheless, the angle of the immediate rotation is one or two orders of magnitude less than the angle in phenomenon (1).

(Using an older device, the strongly dampened oscillations could be observed around the new equilibrium position (i.e. rest-position with twisting tension in the filament) after a mechanical pulse had been given to the plate. Such oscillations were directly observed by Voráček in a very rough experiment performed in March 1991 (not published) and (after a personal report) discovered by Vybíral when reanalyzing observational material from 1978.)

(3) The critical angle of rotation in phenomenon (1) is dependent on the period of time the water has been at rest. This angle increases with time, starting from virtually zero. (The critical angle can reach values of several tens of degrees!)

(4) If the plate is only partially immersed, the critical angle is significantly greater than when it is immersed completely.

(5) In the case of partial immersion, the phenomenon analogous to (2) is much more prominent.

(Both phenomena (4) and (5) are time-dependent as described in (3) despite the time-invariance of the surface tension which we also tested.)

(6) If the water is stirred – not too thoroughly – after having been at rest for several days, then, when being again at rest, the critical angle increases from zero quicker than when new "fresh" water is used.

(7) The critical angle is significantly augmented and the phenomenon appears earlier if the (distilled) water is boiled (thus deaerated) before the experiment is started.

Regarding the convincing prominence of observed phenomena, we consider that they are significant enough in order to claim they are real, despite the performed experiments being only qualitative. Our first attempts in carrying out quantitative experiments with any acceptable precision were not successful. This was due to a too large dispersion of the measured values; much more sophisticated laboratory equipment than we could acquire, as well as more strict measurement conditions than those we could guarantee, were necessary. In spite of many problems, the experiments have since been performed (Vybíral, 2006). At that occasion it was incidentally discovered that the phenomena never appeared in the water which was deionized.

In accordance with the generally accepted terminology, the above described complex of phenomena could appropriately be called the *autothixotropy* of water.

2 Proposed Explanation of the Phenomena

Looking for a preliminary interpretation of our observations, a hypothesis based on "ephemeral polymerisation" of the water seems to be plausible. The existence of such a weak/false polymerisation was suspected decades ago, both defended and denied by experts. If the ephemeral polymerisation of water is the cause of the observed phenomena, it indicates that dipole water molecules are establishing chains or a network; first as minute complexes and thereafter combining successively with one another. The structure then becomes more and more dense while oscillating at a certain amplitude on a scale of molecules. Such a structure will be relatively fragile, susceptible perhaps even to differences in the concentration of materials dissolved in water, at different points inside the vessel. Brownian motion can be observed in the case of a conglomerate of molecules having a non-polar character, owing to collisions with molecules of water oscillating in the established network. A stirring – not too thorough – of the "old" water seems to preserve parts of the network, making the subsequent "dipole polymerisation" quicker than it would be in the "fresh" (i.e. well stirred) water. Further, the established structure has a certain degree of elasticity. If the water is deaerated by boiling, no dissolved air (gases) has a disturbing influence either on the developing process or integrity of the structures; consequently, the phenomenon appears earlier and is more pronounced.

Our observations are consistent with the recently published results of Wernet et al. (2004).

3 Dependence of the Manifestation of Autothixotropy on the Degree of Molecular Translative Freedom and its Possible Influence on the Cytoskeleton

A manifestation of the autothixotropy of water depends – among others – on the degree of freedom of the translative motion of its molecules. The freedom is limited close to the boundary between the water and some other environment, e.g. a solid body or the air-atmosphere over the surface of the water. The freedom of the molecular motion is then limited relatively very deep into the water-body, not only in one molecular layer but, perhaps, on the scale of several hundred multiples of its thickness or more. The limited degree of freedom, depending on the number of free space-dimensions less than three, appears as follows:

(1) In the situation where the free motion is limited to two space-dimensions i.e. – more or less – to a plane, one can find its relevant manifestation in the phenomena (4) and (5) described in the above Section on laboratory experiments.
(2) When a thin capillary tube is used, the free molecular transitive motion is limited in practice to just one dimension, resulting in the apparently undeniable phenomenon of the *polywater*, observed decades ago, at that time claimed to be a sensational discovery, which – however – soon after proved to be false.
(3) When the transitive freedom is limited in all possible directions, i.e. in all three space-dimensions, the manifestation of the autothixotropy must – quite logically – become very prominent and influential. Such a situation occurs in small cellular spaces and possibly significantly influences, or even determines, the rigidity of the cytoskeleton.

Note

Actually two diametrically different results sustained by serious observations exist: According to the first one (Cowan et al., 2005), the clusters in the water have a duration of less than one hundred femtoseconds, while, according to the second one (made by the authors), the clusters are growing to the webs on the time scale of days. Since the webs never arose in the water which was deionized, we believe that the purity of the water can be a decisive factor; we must admit that the distilled water used was not perfectly pure and could have been significantly contaminated by salt ions, even if only to a very minute degree.

From a comparison of experiments with distilled water and deionized distilled water, it is possible to deduce that kernels of macroscopic clusters of water molecules are the salt ions contained in water.

Motto a posteriori: If two different observations seem to be mutually incompatible within the frame of an established theory, the most probable explanation is not that one of the observations must be wrong, but that the theory is wrong or – at least – incomplete, and that the observations merely discovered that it was not self-consistent.

References

Cowan, M. L. et al. (2005) *Nature* **434**, pp. 199–202.
Horák, Z. (1984) *Astrophys. Space Sci.* **100**, pp. 1–11.
Vybíral, B. (1987) *Astrophys. Space Sci.* **138**, pp. 87–98.
Vybíral, B. (1989) In: *Sborník Pedagogické Fakulty* **54**, Fyzika, pp. 307–318, SPN, Praha, (in Czech).
Vybíral, B. (2006) In: G. Pollack et al.(eds.), *Water and the Cell* (Chapter 15, pp. 299–314: *The Comprehensive Experimental Research on the Autothixotropy of Water*), Springer, Dordrecht.
Wernet, Ph. et al. (2004) *Science* **304**, pp. 995–999.

Propagation of Volume Phase Transitions as a Possible Mechanism for Movement in Biological Systems

L. Yeghiazarian and R. Lux

Abstract Scientists frequently use technical terms to describe biological systems and functions. By virtue of technical analogy, these terms provide us with insight into the mechanisms that drive biological systems, and often guide us in exploration of little understood phenomena. In this article, we go from a well understood engineered system to less understood biological ones. We apply our knowledge of polymer gel based devices, to motility principles of two microorganisms: *Vorticellid* ciliates and non-flagellated cyanobacterium Synechococcus. We propose that contraction of *Vorticellid* and swimming of *Synechococcus* are based on the same mechanism that drives the movement of polymer gels, namely the propagating volume phase transitions.

Keywords Gel · polymer · volume phase transition · *Vorticellid* · *Synechococcus* · swimming · contraction mechanism

1 Introduction

The term "biomimetic" is commonly used to describe a technology that is built after a certain biological organism, feature, or function. It implies therefore that the flow of knowledge and inspiration between "bio" and engineering is unidirectional: from "bio" to engineering. While there can be little argument about the truthfulness of this statement, it is worth noting the existence of information and technology flow in the opposite direction, i.e. from engineering to "bio". Let us explore this idea in greater detail.

The flow of knowledge from engineering to "bio" often takes a descriptive shape, with examples abundant in cell biology and micro- and nanotechnology. We commonly use terms such as "engines, locomotives, motors, actuators, sensors, machines, fuel", etc. to describe biological structures such as molecular motors and

L. Yeghiazarian
Department of Bioengineering, currently at the Department of Biostatistics, School of Public Health, UCLA, Los Angeles, CA, USA

ion pumps, and their functionality. Cell membrane processes are routinely expressed in terms such as conductivity, insulation and capacitance, most commonly encountered in electrical engineering. Perhaps the most famous example is the comparison of F_1F_0-ATPase, an enzyme that produces ATP in mitochondria and in the majority of microorganisms, to a rotary engine. We refer the reader to an excellent review by Knoblauch and Peters (Knoblauch and Peters 2004) for an in-depth discussion of these issues.

Even though such comparisons appear on the surface purely descriptive, by virtue of technical analogy they do nonetheless provide us with explanatory interpretation and insight into the actual mechanisms that drive these biological entities. The analogy is therefore not just morphological, but also functional. Taking one step further, these analogies often guide us in our investigation of phenomena that are not fully understood. Our knowledge of the principles that form the basis of engineering creations may eventually lead us to the yet uncovered principles underlying their biological "counterparts".

In step with these ideas, we describe here some of the mechanisms of mobility recently investigated in artificial hydrogels, and explore the possibility that similar mechanisms are responsible for movement in two biological organisms: *Vorticellid* ciliates and the non-flagellated motile cyanobacterium *Synechococcus*.

2 Gels and Volume Phase Transitions

Gels consist of solvent and a cross-linked polymer matrix. If the solvent used is water, they are called hydrogels. The polymer can be of natural (polysaccharides, proteins) or synthetic (poly (vinyl alcohol), poly (hydroxyethyl methacrylate)) origin (Yamauchi 2001).

Apart from being used in desserts (think Jell-O), gels are becoming increasingly popular and extensively investigated in materials science, engineering, biotechnology and pharmaceuticals. One of the most researched properties of gels is their capability to shrink and swell thus changing their volume several hundred-fold in response to changes in environmental stimuli (Fig. 1). As the gel shrinks, the solvent is removed from its polymer matrix; during swelling the gels re-absorb the solvent. Stimuli include temperature, pH, ion concentration etc. Under some conditions, this dramatic change in volume is discontinuous, i.e. the gel swells or shrinks in response to an infinitesimal change at the critical value of an environmental parameter (Fig. 2) (Hirotsu et al. 1987, Matsuo and Tanaka 1988). This property is called a volume phase transition – a phenomenon that might be the mechanism behind motility of some biological organisms, as we will discuss later.

Generally, volume phase transitions occur due to imbalance in (1) polymer-polymer affinity as mediated by the solvent, (2) elasticity of the cross-linked polymer network, and (3) presence of ions driven forces, or pressures, acting within the gel (Hirotsu et al. 1987). Polymer-polymer affinity ultimately stems from mixing between the polymer network and the solvent. These interactions give rise to a force that tends to expand or contract the gel depending on the chemical and electrical

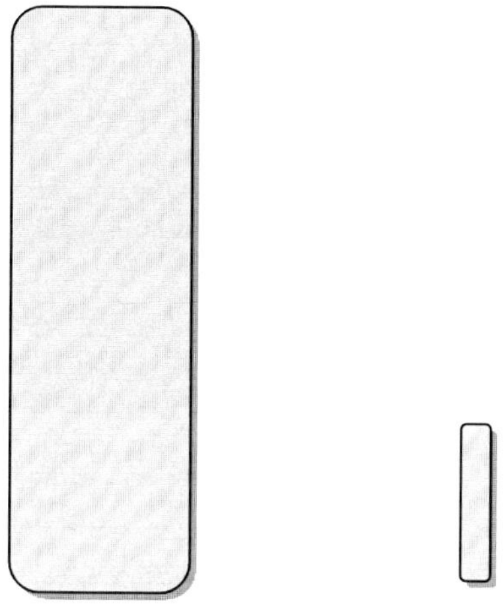

Fig. 1 A drawing of a gel in a swollen and collapsed state. The gel is capable of reversibly changing its volume in response to an infinitesimal change at the critical value of an environmental parameter

properties of the participating molecules. This force is countered by the elasticity of the polymer network. The presence of ions changes the balance of forces depending on interactions of ions with other molecular components of the gel (Tanaka 1981).

As an example of interplay between the three forces, consider thermosensitive gels such as the widely used N-isopropylacrylamide (NIPA) gels. NIPA monomers have hydrophobic side chain groups that are shielded by water cages at lower temperatures. As the temperature is increased, the ordered structure of water molecules

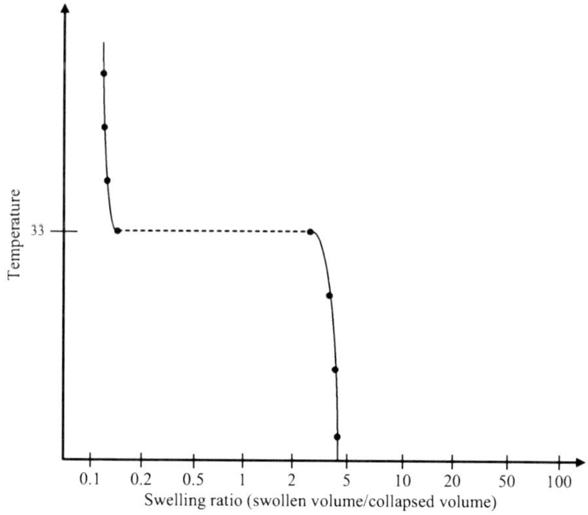

Fig. 2 A graph showing the dependence of the swelling ratio on ambient temperature for a temperature sensitive N-isopropylacrylamide (NIPA) gel (drawing based on work of Hirotsu et al. 1987; Tanaka 1981). The gel is swollen at low temperatures. The volume decreases dramatically at the critical temperature of approximately 33°C, as the gel undergoes a volume phase transition. A discontinuity is observed in the swelling ratio at the critical temperature: it jumps by an order of magnitude. In some cases the volume may change by two or more orders of magnitude

around the gel network breaks, the polymer network aggregates, the gel collapses and the water is removed from the gel (Annaka et al. 2000). NIPA gels swell at temperatures below about the critical 33° C and collapse above it.

Another example of environmentally sensitive gels are ion-responsive hydrogels. These gels contain groups that selectively react with particular ions. As ions bind to these groups, the balance of forces is tilted, and the gel changes its volume accordingly (van der Linden et al. 2003).

We have harvested the capability of NIPA gels to undergo volume phase transitions, in the development of a mobile thermosensitive gel-based device. The movement was initiated and sustained by capitalizing on volume phase transitions controllably induced in the gel (Yeghiazarian et al. 2005).

The device consists of a cylindrical NIPA gel confined in a glass tube. The basic underlying principle of directional movement lies in the propagation of a volume phase transition along the length of the gel. As mentioned before, at room temperature the gel is swollen. As we heat the gel from one end, it collapses. We then move the source of heat along the body of the gel, and as we do so, the gel continues shrinking in the same direction. Once the gel has collapsed, we start cooling it from the same end. It swells and pushes the rest of the gel forward until it swells back to its original volume. It should be mentioned that each segment of the gel swells and

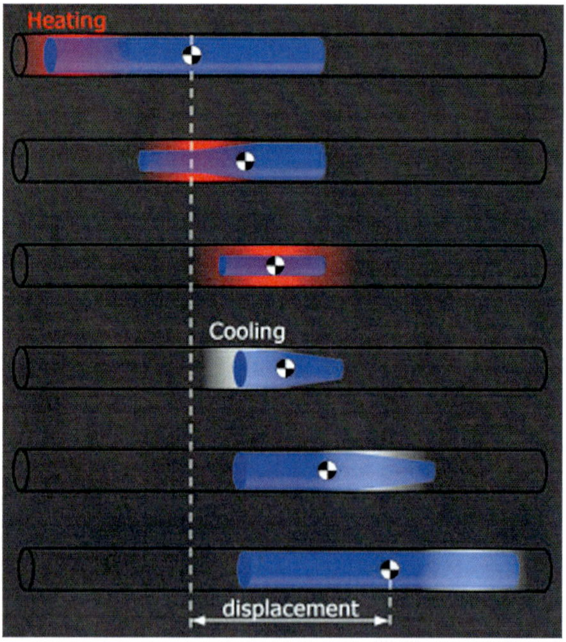

Fig. 3 A schematic of an experiment demonstrating linear displacement of the center of mass of a polymer gel within a glass capillary. It is achieved by propagating a volume phase transition along the body of the gel in the desired direction of movement. Drawing courtesy of Wiesner group, Cornell University

shrinks isotropically, i.e. in equal proportions in all directions. The schematic of our experiment (Fig. 3) shows a significant displacement of the gel after several cycles of heating and cooling.

Gels that move by propagating volume phase transitions can be seen as having waves travel along their bodies. We will come back to the idea of traveling waves when we discuss the movement mechanisms of non-flagellated cyanobacterium *Synechococcus*. We are now ready to explore the motility of our two target organisms, the *Vorticellid* ciliates and the non-flagellated cyanobacterium *Synechococcus*. We will present evidence for the idea that these organisms might be using a mechanism similar to propagating volume phase transitions.

3 Contraction of Spasmoneme in *Vorticellid* Ciliates

Motility in most eukaryotic organisms is associated with hydrolysis of ATP, the universal energy source. There are some organisms however, that are ATP independent yet highly efficient in terms of movement (Knoblauch and Peters 2004). Perhaps some of the best-known examples of such independence are *Vorticellid* ciliates such as *Vorticella, Carchesium*, and *Zoothamnium*. These are single cell organisms consisting of a zooid and a long stalk. They attach themselves to substrates with the stalk (Fig. 4). They have drawn a lot of interest because of the contractile capability of the stalk: it coils in an all-or-nothing fashion in response to environmental hazards such as predators. The stalk coiling is a result of contraction of a rod-shaped intracellular helical organelle, the spasmoneme which resides within the stalk.

In Sections 3.1–3.2 we summarize some of the known characteristics of *Vorticella convallaria* and *Giant zoothamnium* (Moriyama et al. 1998; Moriyama et al. 1999). We comment on these characteristics and discuss hypotheses of contraction mechanisms in Section 3.3.

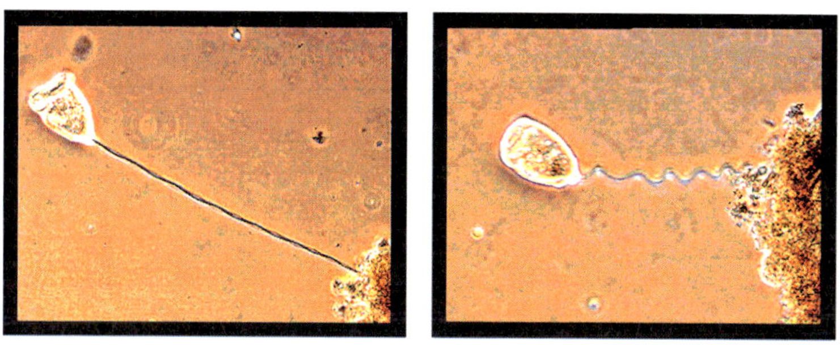

Fig. 4 A phase contrast microscope image of a living Vorticella shown in extended (*left*) and contracted (*right*) state. Image courtesy of Dr. Rick Gillis, *http://www.bioweb.uwlax.edu/zoolab/*

3.1 Spasmoneme Composition

The biochemical features of the spasmoneme reflect the composition and behavior of ion-sensitive gels. Even though the spamoneme appears to contain organelles including endoplasmatic reticulum (Moriyama et al. 1999), the main "building block" of the spasmoneme is spasmin. This 20 kDa protein forms 2–4 nm wide weakly cross-linked filaments that appear to be roughly aligned. Binding of Ca^{2+} ions renders spasmin hydrophobic and results in the collapse of the spasmin network (Mahadevan and Matsudaira 2000) Since the dry volume of the spasmoneme, as well as the number of filaments and crosslinks, do not change in the extended and collapsed state (Moriyama et al. 1999), the structure of the spasmin network may not be altered upon addition or removal of Ca^{2+}. The overall spasmoneme configuration is helical, which causes it to rotate during contraction (Mahadevan and Matsudaira 2000).

3.2 Characteristics of Contraction and Environmental Requirements for Contractility

The contraction is driven by changes in Ca^{2+} concentration in the environment, and is fully reversible. No ATP or pH dependency has been found. Contraction and extension cycles can be repeated many times *in vitro* by simply varying the Ca^{2+} concentration (Moriyama et al. 1999). The critical Ca^{2+} concentration threshold is about 0.2 μM, saturating at 1 μM. While contraction takes only a few milliseconds, the extension time is much longer (a few seconds).

When exposed to Ca^{2+}, the stalk begins coiling near the zooid, and coiling spreads distally (Moriyama et al. 1998). The stalk contracts up to 50% of its original length and diameter, achieving maximum speeds of 8–9 cm/s. The fact that both length and diameter change in same proportion, means that the spasmoneme contracts isotropically, i.e. equally in all directions. The total volume decreases dramatically by up to 86% of the volume in the extended state (Moriyama et al. 1999). The stalk itself in the extended state is about 2–3 mm long. A car moving with equivalent velocity would be traveling at an astonishing 260,000 km/h.

The force of contraction is calculated to be on the order of a millidyne, and the power generated during contraction is several milliergs per second. This makes the spasmoneme one of the most powerful biological engines in terms of specific power per unit mass (Mahadevan and Matsudaira 2000).

3.3 Mechanisms of Contraction Initiation and Propagation

The exact mechanisms of contraction initiation and propagation remain unclear. It does not appear to be caused by a sliding mechanism like in muscles (Moriyama et al. 1999). Muscle contraction is driven by energy liberated from ATP, while the spasmoneme is entirely ATP-independent.

The currently accepted working hypothesis is based on the fact that the contraction is induced by Ca^{2+}, and that spasmoneme contains Ca^{2+} binding proteins. The spasmoneme has been described as a supramolecular spring with rubber-like elasticity that contracts due to imbalance between osmotic and entropic forces, just like in a polyelectrolyte gel. The energy is stored by electrostatic repulsion between the negatively charged filaments. The charges are neutralized in the presence of Ca^{2+}, and the spring collapses entropically (Moriyama et al. 1998; Mahadevan and Matsudaira 2000). Ca^{2+} is hypothesized to be contained in membranous tubules within the spasmoneme, and globally released from these stores. This proposition has not been proven (Moriyama et al. 1998).

We would like to build on this theory by pointing out several similarities in observations of polymer gel and spasmoneme behavior, and by offering inferences based on our working knowledge of gel-based devices.

The facts presented in Sections 3.1–3.2 provide us with very valuable information. It appears that the spasmoneme is a cross-linked network of spasmin proteins. Since the dry volume of the spasmoneme does not change as it swells and collapses, the change in volume can be attributed to the solvent entering and leaving the spasmin network. It is also reasonable to assume that since the number of filaments and cross-links remains constant throughout, the network structure doesn't change, and a polymerization/de-polymerization mechanism (Mogilner and Oster 1996) can be excluded.

We also note that the spasmin network is responsive to the presence of Ca^{2+} ions, which indicates that we are dealing with an ion-sensitive system. It is known that spasmin contains groups that become hydrophobic when bound to Ca^{2+} ions. Moreover, the system undergoes critical changes as the concentration of Ca^{2+} ions crosses over a certain threshold, in this case 0.2 μM. The result of this crossover is a discontinuous change in the spasmoneme volume. In the light of our knowledge of polymer gel behavior, we could speculate that the spasmoneme is actually an ion-sensitive protein gel capable of volume phase transitions.

Let us now look into a seemingly minor observation of difference between the contraction and extension times of the spasmoneme. The contraction time is significantly shorter than the extension time. Such discrepancy is typical of polymer gels, which rely on diffusion to change their volume. Kinetics of shrinking and swelling are quite complex, and have been explored in detail by Matsuo and Tanaka (Matsuo and Tanaka 1988). Here it is sufficient to say that there is a power law relationship between the time required to resume the original volume and the gel radius, with different exponents for shrinking and swelling. As a rough approximation however, the time is considered proportional to the squared gel radius (Tanaka et al. 1980). An interesting consequence of this relationship is that the time it takes the gel to change its volume decreases dramatically as the gel radius gets smaller. For example, if 1 cm diameter gel needed 36 days to swell, a gel of 1 μm radius would swell in 3 seconds (van der Linden et al. 2003). This acceleration of time constants on smaller scales may account for the large speed with which the spasmoneme collapses that we mentioned in Section 3.2. Exact calculations of time constants and attainable velocities (Matsuo and Tanaka 1988; Yeghiazarian et al. 2005) require knowledge of several spasmoneme parameters, in particular the collective diffusion coefficient, that are not currently available.

Next, recall that the contraction originates near the zooid and then spreads along the stalk. This might be an indicator that spasmoneme coiling is due to a propagation of a volume phase transition along its length. This also indicates that Ca^{2+} may not necessarily be released simultaneously from all storage locations. If Ca^{2+} were released globally at the same time, the entire body of the stalk would coil simultaneously. That has not been observed.

Finally, let us turn towards the last two issues: the reversibility and the isotropic nature of volume change. In polymer gels, volume changes due to phase transitions have been shown to be fully reversible in response to a change in the environmental conditions. The same is true about reversibility of spasmoneme volume in response to concentration of Ca^{2+}.

As we have mentioned earlier, polymer gels exhibit isotropic swelling and shrinking. This means that a segment, or an entire gel, whichever is subjected to an environmental stimulation, will change its volume in the same proportion in all directions. For example, both length and diameter will shrink or swell by 50%. Propagation of volume changes is then possible by exposing individual segments of the gel sequentially and controllably to stimulation, however each individual segment would swell/shrink isotropically. When the entire gel is swollen or collapsed, the proportional change from its original dimensions is the same in all directions. Since we know that the final volume change in the spasmoneme is isotropic, and that the coiling begins at a certain point and then spreads along the stalk, we could conclude that the spasmoneme might be undergoing a traveling volume phase transition.

To wrap up the discussion of spasmonemes, the initiation mechanism is unknown. Possibly in response to presence of a predator, Ca^{2+} ions are locally released into the spasmin network at the point closest to the zooid. Ca^{2+} ions bind to spasmin strands and render them hydrophobic. Water begins escaping the spasmoneme, the volume starts decreasing near the zooid and the stalk starts coiling. Contraction is induced, and as more Ca^{2+} is being released, the contraction travels along the length of the stalk.

Next, we look into the motility mechanism of the non-flagellated cyanobacterium *Synechococcus*. As we will see, even though *Synechococcus* motility is seemingly very different from the contractile properties of the *Vorticellid* spasmoneme, we will propose that propagating volume phase transitions may be the mechanism used by *Synechococcus* to swim.

4 Swimming Capability of the Non-Flagellated Cyanobacterium *Synechococcus*

We will first present the known facts about the cyanobacterium *Synechococcus* (Section 4.1). We will follow with existing hypotheses on its motility mechanism (Section 4.2), and propose that volume phase transitions could be implicated in the swimming capability of this organism (Section 4.3).

4.1 Structure and Characteristics

The cyanobacterium *Synechococcus* is a unicellular rod-shaped bacterium that swims without the aid of flagella or any other visible swimming organelles. The mechanism of motility of this organism is still unclear. *Synechococcus* is about 0.7–0.9 μm in diameter and 1.25–2.5 μm in length (Waterbury et al. 1985). The cells swim in relatively straight paths, reaching velocities of about 25 μm/s. Linear translocation is accompanied by cell rotation. The speed seems to be independent of light intensity, ATP concentration, or electric field (Pitta and Berg 1995).

It was discovered that motility was not possible in the absence of sodium and calcium ions (Willey et al. 1987; Pitta et al. 1997). The cells stopped moving once the external sodium concentration dropped below 10 mM. Sodium is proposed to be the most probable energy source for motility, as changes in cell velocity correlated with changes in sodium concentration. This indicated that the conversion of energy to produce mechanical motion took place in the cytoplasmic membrane of the bacterium, as is the case with flagellated organisms.

It was also found that motility was dependent on the presence of Ca^{2+} ions, and no divalent atoms could replace Ca^{2+}. The threshold concentration for $Ca^{2+}$was found to be 1 mM, below which motility was arrested.

Prior to the confirmation of an essential role of Ca^{2+} in swimming of *Synechococcus*, a remarkable Ca^{2+}–binding, negatively charged 130 kDa polypeptide protein SwmA that covers the entire cell surface, was discovered. When the external layer containing SwmA was removed, the organism lost its ability to propel itself linearly, however it was still able to rotate (Brahamsha 1996). It was hypothesized that the generation of torque and thrust was independent. It was further found that SwmA formed a layer, called the S-layer, covering the cell. The S-layer was absent in mutated non-swimming cells. Moreover, the presence of the S-layer was the only structural difference between the swimming and non-swimming cells (McCarren et al. 2005).

Some *Synechococcus* strains possess an interesting structural detail on the cell surface, namely short fibrils, or spicules, that extended up to 150 nm into the surrounding liquid, and also inwards towards the inner cell membrane (Samuel et al. 2001). The discovering team suggested that the spicules might be part of the cell's motility mechanism. One possibility was that any oscillations of the cell membrane could be transmitted to the spicules resulting in a rowing motion. Interestingly, experiments with various strains of *Synechococcus* later revealed the existence of wild-type cells without spicules, that were capable of swimming. These results suggested that spicules in general were not required for motility (McCarren et al. 2005).

4.2 Traveling Waves as a Possible Motility Mechanism

In search of a mechanism responsible for swimming of *Synechococcus* and intrigued by the absence of any detectable motility structures, researchers turned to mechanics of deformations. The theory of self-propulsion of deformable bodies in liquids has been around for quite some time. Miloh and Galper calculated sequences of geometric shapes that would sustain self-propulsion for general bodies (Miloh and Galper 1993). It was later proposed that *Synechococcus* might swim by using traveling surface waves. In order to attain the actual velocities of about 25 μm/s, the organism would have to propagate waves of 0.02 μm in amplitude, 0.2 μm in length and 160 μm/s speed (Ehlers et al. 1996; Stone and Samuel 1996).

Pitta and colleagues (Pitta et al. 1997) suggested that local swelling of the cell envelope propagating down the cell body might be responsible for traveling waves. They also indicated that the waves with parameters derived by Ehlers, Stone and colleagues (Ehlers et al. 1996; Stone and Samuel 1996) would not be detectable under a standard phase-contrast microscope. Indeed, no visible deformations have been reported in the literature so far.

Oster and his group proposed that undulations of the cell body could be caused by local bending of postulated elastic mechanochemical filaments located at the inner membrane of the cell. They incorporated the discovery of spicules in some *Synechococcus* strains into their theory and proposed that if spicules are attached to the filaments, then filament bending would make the spicules move and induce cilia-like undulations (Wolgemuth et al. 2003).

We believe that swimming by body deformation in the form of surface waves is a plausible hypothesis for *Synechococcus* motility. Local swelling as a motility mechanism has been proposed by Koch for gliding microorganisms (Koch 1990). Pitta and colleagues (Pitta et al. 1997) suggested that this idea should be considered among others. Here we hypothesize that traveling waves might be due to volume phase transitions propagating along the cell surface.

While *Synechococcus* is much less studied than *Vorticellids*, many of the factors necessary for sustained volume phase transitions are present nonetheless. We observe an all-or-nothing behavior with respect to Ca^{2+} ions. A Ca^{2+}-binding protein layer (the S-layer) with alternating hydrophobic and hydrophilic regions has been found to cover the entire cell surface, and the presence of the S-layer is the only difference between the swimming wild-type and non-swimming mutants discovered so far. In a possible scenario, transient conformational changes of a cell surface protein (SwmA) could be triggered by sequential binding of Ca^{2+}. The sequence and propagation of these conformational changes could be regulated by controlled release of intracellular Ca^{2+} through specific channels along the cell body. The conformational changes of SwmA could lead to a localized decrease of volume, and therefore a local shape change of the cell. The localized volume change would travel along the cell body thus generating traveling waves. If Ca^{2+} is found to render SwmA hydrophobic, then in a somewhat modified scenario the shape change could be induced by a local decrease in volume and therefore a local contraction of the cell wall.

The all-or-nothing characteristic of the cell motility with respect to Ca^{2+} ions indicates that the localized volume change is likely to be due to a discontinuous volume phase transition. As we have mentioned earlier, discontinuity is exhibited in an abrupt change following an infinitesimal variation of the controlling environmental parameter, in this case Ca^{2+} concentration.

We would like to reiterate that this microorganism is not well studied. Unlike *Vorticellids*, *Synechococcus* does not have easily identifiable motility organelles that could be isolated and studied in greater detail and under various conditions. Much of the available information lends itself only to theoretical hypotheses that have not yet been experimentally verified. Until more solid experimental evidence emerges, the theory of traveling waves remains the only plausible explanation of the swimming capabilities of this amazing microorganism.

5 Conclusions

We began this piece with a discussion of analogies between the engineering and biological worlds. While examples of biomimetically engineered structures and devices are abundant, scientists are now more and more often using technical terms to describe biological systems. This indicates that the flow of ideas goes not only from the biological world to the engineering one, but in the opposite direction as well. A very interesting feature of this trend is that by applying our technical knowledge to biological objects and phenomena, we are actually going from much better understood systems towards the less understood ones. By finding similarities in the way technical and biological systems behave, we hope to enhance our explorations of biological systems with engineering knowledge.

In the spirit of these ideas, we looked at the motility mechanisms of two groups of organisms – *Vorticellids* and *Synechococcus*. Based on our recent work on polymer gel-based devices and using our knowledge of propagating volume phase transitions as a mechanism to drive synthetic polymer gels, we proposed that the same mechanism could be responsible for contractions of *Vorticellids* and swimming capability of*Synechococcus*. While information currently available is very much in accord with the propagating volume phase transitions being the actual mechanism behind the motility of these microorganisms, more experimental evidence is needed to confirm or disprove this hypothesis.

Acknowledgments The authors would like to thank Ulrich Wiesner and Claude Cohen of Cornell University for stimulating discussions and critical reading of the manuscript.

References

Annaka, M., K. Motokawa, S. Sasaki, T. Nakahira, H. Kawasaki, H. Maeda, and Y. Tominaga. 2000. Salt-induced volume phase transition of poly (N-isopropylacrylamide) gel. J. Chem. Phys. **113**:5980.

Brahamsha, B. 1996. An abundant cell-surface polypeptide is required for swimming by the non-flagellated marine cyanobacterium Synechococcus. Proceedings of the National Academy of Sciences of the United States of America **93**:6504–6509.

Ehlers, K. M., A. D. T. Samuel, H. C. Berg, and R. Montgomery. 1996. Do cyanobacteria swim using traveling surface waves? Proceedings of the National Academy of Sciences of the United States of America **93**:8340–8343.

Hirotsu, S., Y. Hirokawa, and T. Tanaka. 1987. Volume-phase transitions of ionized N-isopropylacrylamide gels. Journal of Chemical Physics **87**:1392–1395.

Knoblauch, M., and W. S. Peters. 2004. Biomimetic actuators: where technology and cell biology merge. Cellular and Molecular Life Sciences **61**:2497–2509.

Koch, A. L. 1990. The Sacculus Contraction Expansion Model for Gliding Motility. Journal of Theoretical Biology **142**:95–112.

Mahadevan, L., and P. Matsudaira. 2000. Motility powered by supramolecular springs and ratchets. Science **288**:95–99.

Matsuo, E. S., and T. Tanaka. 1988. Kinetics of discontinuous volume-phase transitions of gels. Journal of Chemical Physics **89**:1695–1703.

McCarren, J., J. Heuser, R. Roth, N. Yamada, M. Martone, and B. Brahamsha. 2005. Inactivation of swmA results in the loss of an outer cell layer in a swimming Synechococcus strain. Journal of Bacteriology **187**:224–230.

Miloh, T., and A. Galper. 1993. Self-Propulsion of General Deformable Shapes in a Perfect Fluid. Proceedings of the Royal Society of London Series a-Mathematical Physical and Engineering Sciences **442**:273–299.

Mogilner, A., and G. Oster. 1996. Cell motility driven by actin polymerization. Biophysical Journal **71**:3030–3045.

Moriyama, Y., S. Hiyama, and H. Asai. 1998. High-speed video cinematographic demonstration of stalk and zooid contraction of Vorticella convallaria. Biophysical Journal **74**:487–491.

Moriyama, Y., H. Okamoto, and H. Asai. 1999. Rubber-like elasticity and volume changes in the isolated spasmoneme of giant Zoothamnium sp. under Ca^{2+}-induced contraction. Biophysical Journal **76**:993–1000.

Pitta, T. P., and H. C. Berg. 1995. Self-Electrophoresis Is Not the Mechanism for Motility in Swimming Cyanobacteria. Journal of Bacteriology **177**:5701–5703.

Pitta, T. P., E. E. Sherwood, A. M. Kobel, and H. C. Berg. 1997. Calcium is required for swimming by the nonflagellated cyanobacterium Synechococcus strain WH8113. Journal of Bacteriology **179**:2524–2528.

Samuel, A. D. T., J. D. Petersen, and T. S. Reese. 2001. Envelope structure of Synechococcus sp. WH8113, a nonflagellated swimming syanobacterium. BMC Microbiology **1**.

Stone, H. A., and A. D. T. Samuel. 1996. Propulsion of microorganisms by surface distortions. Physical Review Letters **77**:4102–4104.

Tanaka, T. 1981. Gels Scientific American **244**:124–137.

Tanaka, T., D. Fillmore, S.-T. Sun, I. Nishio, G. Swislow, and A. Shah. 1980. Phase transtions in ionic gels. Physical Review Letters **45**:1636–1639.

van der Linden, H. J., S. Herber, W. Olthuis, and P. Bergveld. 2003. Stimulus-sensitive hydrogels and their applications in chemical (micro) analysis. Analyst **128**:325–331.

Waterbury, J. B., J. M. Willey, D. G. Franks, F. W. Valois, and S. W. Watson. 1985. A Cyanobacterium Capable of Swimming Motility. Science **230**:74–76.

Willey, J. M., J. B. Waterbury, and E. P. Greenberg. 1987. Sodium-Coupled Motility in a Swimming Cyanobacterium. Journal of Bacteriology **169**:3429–3434.

Wolgemuth, C. W., O. Igoshin, and G. Oster. 2003. The motility of mollicutes. Biophysical Journal **85**:828–842.

Yamauchi. 2001. Gels: introduction. In Gels handbook, Y. Osada (Ed.). Academic Press.

Yeghiazarian, L. L., S. Mahajan, C. D. Montemagno, C. Cohen, and U. Wiesner. 2005. Directed motion and cargo transport through propagation of polymer-gel volume phase transitions. Advanced Materials **17**:1869–1873.

Cell Plasma Membranes and Phase Transitions

Mark M. Banaszak Holl

Abstract Cell plasma membrane phase plays a large role in membrane trafficking and signaling. The role that membrane phase plays in cell function, current hypotheses concerning the size and time duration of the phase transitions, and the role in disease are discussed.

Keywords Cell · plasma membrane · lipid raft · gel · fluid mosaic model

Cellular plasma membranes consist of a variety of molecules including lipids, cholesterol, and protein. For decades, it has been noted that these materials exhibit liquid crystalline behavior and that the combination of oriented structure, flow, and lability are important characteristics for optimal functioning of the plasma membrane. (Bernal, 1933; Stewart, 1961) Indeed, these are the key physical characteristics which were considered in the development of the *Fluid Mosiac Model* of the cell plasma membrane. (Singer and Nicolson, 1972) In addition to oriented structure, flow, and lability, liquid crystalline cell plasma membranes have another intriguing property. Namely, the membranes are often just above the temperature for transition to the gel phase. This phase transition is likely of great significance in the homeostasis of raft structures in the cell membrane. (Simons and Ikonen, 1997; Vereb et al., 2003) Lipid raft structures, localized regions enriched with species such as cholesterol, glycolipids, and sphingolipids, (Simons and Ikonen, 1997; Simons, and Vaz, 2004) are proposed to play important roles in a great variety of membrane trafficking and signaling events. In addition, the phase transition of the plasma membrane from a majority liquid crystalline state to a majority or completely gel state may play an important role in suspending all membrane trafficking activities. The subtle interplay between membrane fluid and gel was succinctly summarized by Mouritsen as follows:

M.M. Banaszak Holl
Department of Chemistry and Program in Macromolecular Science and Engineering, Michigan Nanotechnology Institute for Medicine and Biological Science, 930 N. University Avenue, University of Michigan, Ann Arbor, MI 48109-1055

G.H. Pollack, W.-C. Chin (eds.), *Phase Transitions in Cell Biology*,

> "Biological activity almost universally requires that membranes be in their fluid state; in many cases, just above the transition point" (Mouritsen, 1987)

Although the "transition point" described by Mouritsen may in fact be rather broad for a cell membrane (tens of degrees), it is well known that gel phase membranes exhibit dramatically decreased permeability and that the rates of active transport events such as endo or exocytosis decrease dramatically and frequently cease altogether. (Chapman, 1975; McElhaney, 1984) The evidence that the membrane phase transitions are important has been gathering for over 70 years. However, papers are still frequently published that ignore the importance and/or the existence of membrane phase transition on membrane trafficking events. It remains common for authors to focus on enzymatic cascades.

The texts used to train generations of biochemists and cell biologists provide an interesting lens through which to examine the evolution of understanding regarding the role of cell plasma membrane phase transitions. For example, the 2nd Edition of Biochemistry by Stryer, published in 1981, lists seven common features of biological membranes. (Stryer, 1981) Phase transitions are not mentioned and it is stressed that the membranes are fluid structures. The 2nd Edition of Biochemistry by Voet and Voet published in 1995 has a list of key properties of lipid aggregates. (Voet and Voet, 1995) Once again it is stressed that lipid bilayers are two-dimensional fluids although there is also section describing the variation of fluidity as a function of temperature and the phase change to the gel state is now explicitly mentioned. However, the focus remains on how fluidity allows activity not on how phase transitions might serve as a biological mechanism for modulating action. The 4th Edition of Molecular Biology of the Cell by Alberts et al. published in 2002 also highlights both the fluidity as well as the temperature and composition dependence of the fluidity. (Alberts et al., 2002) The focus is on how cells can maintain a fluid structure at a variety of temperatures by varying membrane composition. Once again the potential role of global phase transitions in modulating overall behavior is not addressed. However, the role of lipid rafts (local changes in membrane composition) in biological activity is introduced. The 3rd Edition of Biochemistry by Hames and Hooper published in 2005 focuses on the fluid mosaic model and also includes explicit discussion of the fluid to gel phase transition, the role of cholesterol, and the importance of lipid rafts in cell signaling. (Hames and Hooper, 2005) In summary, using textbooks as a window on our understanding of membrane function, we don't see a role proposed for the transition of an entire cell membrane from liquid to gel state but do find discussions in terms of localized gel phase regions. A detailed discussion of how regulation and homeostasis of phase (e.g. lipid rafts or domains) modulates biological activity is still not found.

The textbooks reflect the lack of consensus in the tertiary scientific literature regarding the variety of roles which cell plasma membranes may play. However, a survey of the primary scientific literature yields many original scientific studies that bear on these questions. In addition, the secondary literature provides a number of review articles and book chapters. The remainder of this review will focus on generalized models of membrane structure and a number of key experimental cases where substantial evidence regarding the role of cell plasma phase transitions in cell function has been discovered.

1 Lipid Domains or Rafts and Cell Membrane Structure – Updates to the Singer-Nicolson Model

The Fluid Mosaic Model of cell membranes proposed by Singer and Nicolson in 1972 had a dramatic impact upon the scientific understanding of cell membranes. (Singer and Nicolson, 1972) They described the cell membrane by saying

> "In particular, the mosaic appears to be a fluid or dynamic one and, for many purposes, is best thought of as a two-dimensional oriented viscous solution."

Two general consequences of this model were stressed: (1) long range order on the order of 100–200 nm did not typically exist (2) under physiological conditions functional cell membranes would be in fluid state rather than in crystalline (gelled) state. The general principles of this model have served the biological and biochemical communities well in interpreting many important properties of cell membranes. However, an increasing number of experimental observations have suggested that parts of the membrane are not in the fluid state and that phase transitions may play an important role. An interesting comment in the original Singer and Nicolson report pointed to such possibilities when they discussed the broad temperature interval of the liquid to gel phase transition and noted that the breadth suggests only $\sim$100 lipids are involved in the cooperative unit. With hindsight, this value could be taken to suggest a raft domain size of $\sim$30 nm^2 given an average lipid size of $\sim$60 Å^2 and assuming the cooperative unit is bilayer in nature. (Klauda et al., 2006) This estimate suggests a larger raft domain if the protein components of the membrane are also taken into account (up to $\sim$60 nm^2 since the membrane is roughly 50% protein) or if the domain measured by calorimetry is monolayer in nature (once again $\sim$60 nm^2). Based upon the experimental evidence available at the time, and upon additional evidence that has accumulated over the last 35 years, the *Fluid Mosaic Model* is best considered an important limiting case of cell membrane behavior.

At the same time Singer and Nicholson were developing the *Fluid Mosaic Model* another researcher, Chapman, was performing a series of experiments that focused on the phase transition characteristics of cell membranes. His group's work, and the efforts of many other groups, was summarized in a major review published in 1975. (Chapman, 1975) The remarkable correspondence between the phase transition temperature of model lipid bilayer systems and that observed for cells was noted (as it had been in the Singer/Nicolson paper). The review discussed a wide variety of experimental data and ended with a section discussing the significance of lipid phase transitions for biological systems. It was noted that modulation of membrane fluidity, or in other words the extent of gelled membrane regions, would affect a number of important cell characteristics including membrane permeability and signaling. The article ended with a prescient cautionary note regarding applications of cryobiology and the potential impact of lipid phase transitions upon the data obtained.

Papers published in this millennium indicate that the concept of lipid membrane domains, or lipid rafts, has gained wide acceptance. (Edidin, 2003; Jacobson et al., 2007; Kusumi et al., 2005; Lommerse et al., 2004; Mayor and Rao, 2004; Mukherjee and Maxfield, 2004; Munro, 2003; Parton and Hancock, 2004; Simons, and Vaz, 2004) In 2003, a major review article addressed the role of lipid rafts, and the resulting compartmentalization of membrane components, and the impact of these ideas on the *Fluid Mosaic Model*. (Vereb et al., 2003) Single particle tracking experiments, (Edidin et al., 1994; Kusumi et al., 1993) electron microscopy, (Damjanovich et al., 1995) confocal laser-scanning microscopy, (Vereb et al., 2000) and other techniques (Ishitsuka et al., 2005; Miersch and Mutus, 2007) suggest domains on the order of 30,000–500,000 nm^2. Photonic force microscopy experiments suggest raft sizes of $< 2000\ nm^2$ that form aggregates to generate the larger domains. (Pralle et al., 2000) Note that all of these estimates are substantially larger than the size of the cooperativity domain for pure lipid membranes ($\sim$30–60 nm^2) predicted by the breadth of the fluid/gel phase transition as noted in the Singer-Nicolson paper. This is consistent with the view that lipid rafts contain a variety of components including cholesterol, glycolipids, sphingolipids, and protein and the lipid rafts are not simply the cooperativity domains present in pure lipid systems. Estimates of the lifetime of these dynamically restructured domains is estimated to be the on the order of tens of seconds (Dietrich et al., 2002) with fast exchange occurring between the gel and fluid phase domains. The review concludes that an update to the Singer-Nicolson paradigm is needed and the new model should be a "dynamically structured mosaic model." In other words, the Singer-Nicolson model needs the specific inclusion of dynamic phase transitions of the cell membrane to provide the best explanation of the currently available experimental data.

2 Global Cell Plasma Membrane Phase Change – When Might the Entire Cell Plasma Membrane Enter the Gel State?

The importance of a fluid phase membrane for cell function is a central dogma of structural cell biology. (Alberts et al., 2002; Vereb et al., 2003) However, might the global transition of the cell plasma membrane to the gel state serve a useful biological function? This could provide an interesting cellular "switch" for important membrane properties such as permeability, receptor clustering, or changes in surface area required by changes in cell volume. One event during which such a global switch might be useful is apoptosis or programmed cell death. (Bowen et al., 1998) Early in this process, the cytoplasm appears to undergo a gel phase transition in which the volume of the cell decreases by as much as $\sim$45%. (Hessler et al., 2005) Typically termed apoptotic volume decrease or AVD, water is expelled and the concentration of protein in the cytosol increases dramatically. (Bortner and Cidlowski,

1998; Bortner and Cidlowski, 2002; Lang et al., 2000; Maeno et al., 2000; Okada and Maeno, 2001; Vu et al., 2001; Yu et al., 2001; Yu and Choi, 2000) A volume change of this magnitude requires a dramatic ~25% change in surface area of the cell. This could be accomplished by incorporating the cell membrane components into the cell (further increasing the concentration) or releasing the materials to the surroundings (avoided during apoptosis to avoid inducing inflammation). Another possibility is the cell plasma membrane undergoing a liquid to gel phase transition. The lipids occupy 60 Å^2 in liquid phase membrane and only 45 Å^2 in a gel phase membrane. (Klauda et al., 2006) Thus, a cell plasma membrane phase transition can accommodate a ~25% change in cell surface area, roughly what is required for an AVD of ~45%. Later steps in the apoptosis cascade, such as phosphatidyl serine (PS) flipping from the inner to outer leaflet and bleb formation, would seem to require a fluid membrane, or at least domains of fluidity. It is interesting to note along these lines that sphingomyelin depletion concomitant with cholesterol efflux has been proposed to help fluidize the membrane prior to PS slipping. (Tepper et al., 2000)

3 Local Cell Plasma Membrane Phase Change

The importance of the topic of phase transitions in cell membranes is partially indicated by the > 3000 articles published mentioning "lipid rafts" since 1997 (Fig. 1). This particular aspect of phase transitions in cell membranes has garnered substantial attention because of the role the lipid rafts play in cell function, particularly signal transduction and material transport across the membrane. The combination of nanometer size, time dynamics, and cell complexity and diversity has kept the study of lateral membrane phase heterogeneities a challenging endeavor. It remains a key challenge in the nascent field of nanobiology.

Local phase changes in cellular plasma membrane are believed to be important for a great variety of cell functions (Blazyk and Steim, 1972; Brown and London, 1998) Evidence for the importance of local membrane phase for trafficking materials in and out of the cell is substantial. A term in common parlance for these local regions is "lipid raft." It is proposed that raft mediated mechanisms play an

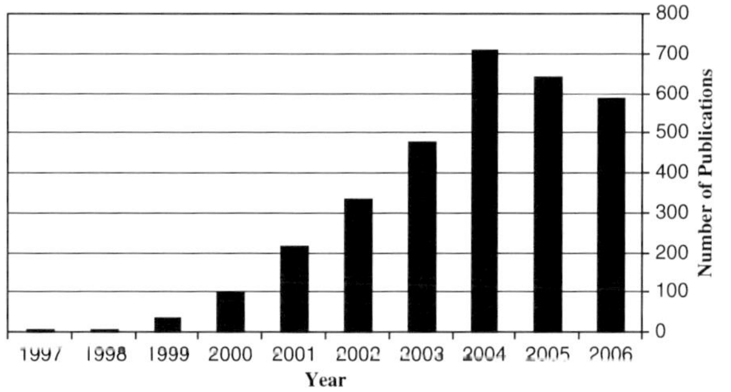

Fig. 1 Publications on the topic of lipid rafts 1997–2006

important role for some forms of endocytosis. (Allen et al., 2007; Hammond et al., 2005; Jacobson et al., 2007; Munro, 2003; Nichols, 2003; Parton and Richards, 2003; Pelkmans et al., 2004; Simons and Ikonen, 1997) Caveolin may also serve to stabilize ~50 nm domains (Anderson et al., 1992; Pelkmans et al., 2004; Rejman et al., 2004) and this may also be a gelled region. Lipid rafts are also proposed to play an important role in great variety of cell signaling events. (Callera et al., 2007; Freeman et al., 2005; Holowka et al., 2005; Karnell and Monroe, 2006; McFarland et al., 2006; Plowman and Hancock, 2005; Simons and Toomre, 2000; Yang et al., 2004) Immunoglobulin E (IgE), T-cell antigen receptor (TCR), glial-cell-derived neurotrophic factor (GDNF), Ras, and Hedgehog signaling have all had raft-based mechanisms proposed. (Simons and Toomre, 2000) Evidence regarding lipid rafts in intracellular membranes is scant and there are far fewer reports to date relating lipid rafts to exocytosis events. (Chintagari et al., 2006; Musch et al., 2004; Salaun et al., 2004; Xia et al., 2004) However, a number of reports have recently appeared connecting SNARE protein and exocytois with lipid rafts. (Kay et al., 2006; Puri and Roche, 2002; Puri and Roche, 2006; Salaun et al., 2005) Lipid rafts are also believed to play important roles in cell adhesion and motility.

4 Biological Systems as Testing Grounds for Phase Hypotheses

Pure lipid bilayers, whether in the form of vesicles or on supports such as glass or mica, are convenient models that are easily formed. Bilayers containing cholesterol, sphingolipids, and/or protein can also be constructed. These have well defined phase transition characteristics that can be quantitatively measured. However, there is concern that these oversimplified systems may yield too simplistic a picture of the membrane behavior of living cells. Researchers employing these convenient biophysical models need to keep in mind the admonition attributed to Einstein, "Everything should be made as simple as possible, but not simpler." One method to simplify the complex problem of cell membrane behavior is the selection of simpler organisms. In this regard, *Acholeplasma laidlawii* has been studied in great detail. (McElhaney, 1984)

Acholeoplasma laidlawii (*A. laidlawii*) belongs to a group of fermentive mycoplasmas that are readily grown in the laboratory. These prokaryotic cells lack a cell wall and internal membranes. The cell plasma membrane therefore contains the vast majority of the cell's lipid and much of the cellular protein. The cells are readily lysed to yield the plasma membranes alone. (Razin, 1979) This system has been amenable to a wide variety of techniques including differential scanning calorimetry and differential thermal analysis, X-ray diffraction, nuclear magnetic resonance, electron spin resonance, infrared spectroscopy, (Cameron et al., 1985) and light scattering. The cells were used for the seminal studies that demonstrated gel-to-liquid lipid phase transitions occur for living *A. laidlawii* cells as well as the isolated membranes. (Melchior et al., 1970; Reinert and Steim, 1970; Steim et al., 1969) Exposure of the cells to exogenous lipid allows variation of plasma membrane lipid content with a concomitant change in the gel-to-liquid phase transition temperature. (McElhaney, 1974a,b; Silvius and McElhaney, 1978) Furthermore, it was found that the cells could be successfully grown at a variety of temperatures provided that the lipid composition was varied so as to keep the liquid-to-gel phase transition just below the growth temperature.

An interesting variant of the relationship of temperature and plasma membrane lipid composition was carried out by analyzing the hypothalamus and brain of garden lizards, *Calotes versicolor*, acclimatized to 16, 26, and 36°C for a period of thirty days. (Durairaj and Vijayakumar, 1984) The lizards provide an interesting test animal because they have no physiological thermoregulation but must control membrane properties by varying membrane composition. The lizards exhibited dramatic changes in the concentration of major phospholipids such as phosphatidylcholine, phosphatidylethanolamine, phosphatidylinositol, cardiolipin, and sphingomyelin. Temperature acclimation has been observed to vary the lipid composition of carp muscle microsomes (Yang and Endo, 1994) and erythrocytes. (Dey and Farkas, 1992) Similarly, a season variation has been noted in the lipid composition in the brains of Rainbow trout (*Clarias batrachus*). (Roy et al., 1997)

5 Hypothesized Roles of Membrane Phase in Disease

In 1785, Benjamin Rush reported that alcohol consumption reduced immunity to infection. (Rush, 1785) Subsequent detailed studies have confirmed this observation. Recent reports indicate that one pathway of ethanol action is disruption of lipid rafts and that this directly modulates immune function. (Szabo et al., 2007) Lipid rafts have been implicated in prion disease as the location of the post-translational conformation conversion of the normal prion protein to the infectious form. (Taylor and Hooper, 2006) Lipid rafts may play a role in Alzheimer's disease as evidence suggest that amyloid precursor protein processing may occur at rafts and that the rafts may be involved in the aggregation of amyloid-ß peptide to form placques. (Cordy et al., 2006) The infection of human erythrocytes by the malarial parasite causes membrane sorting and signaling events that rely on proteins present in lipid rafts. (Murphy et al., 2006) Defective or faulty signaling associated with lipid rafts has also been hypothesized as playing a role in prostate cancer (Freeman et al., 2005) and in hypertension. (Callera et al., 2007)

6 Summary

Phase transitions play a major role in cell plasma membrane function. Variation of overall plasma membrane composition, frequently just above the liquid-to-gel phase transition, is an important whole membrane property for many cells. In addition, control over the phase of localized patches of membrane is important for key cell functions involving internalization and excretion of material and signaling.

Continued efforts to understand the interplay between membrane phase and enzymatic activity will lead to a better understanding of biology and greater insight into disease mechanisms.

References

Alberts B, Johnson A, Lewis J, Raff M, Roberts K, and Walter P. 2002. *Molecular Biology of the Cell*. Garland Science, New York.

Allen JA, Halverson-Tamboli RA, and Rasenick MM. 2007. Lipid raft microdomains and neurotransmitter signalling. *Nature Reviews Neuroscience* **8**: 128–140.

Anderson RGW, Kamen BA, Rothberg KG, and Lacey SW. 1992. Potocytosis: Sequestration and transport of small molecules by Caveolae. *Science* **255**: 410–411.

Bernal JD. 1933. Liquid crystals and anisotropic melts. A general discussion. *Transactions of the Faraday Society* **29**: 1082.

Blazyk JF, and Steim JM. 1972. Phase-Transitions in Mammalian Membranes. *Biochimica Et Biophysica Acta* **266**: 737–741.

Bortner CD, and Cidlowski JA. 1998. A necessary role for cell shrinkage in apoptosis. *Biochemical Pharmacology* **56**: 1549–1559.

Bortner CD, and Cidlowski JA. 2002. Apoptotic volume decrease and the incredible shrinking cell. *Cell Death and Differentiation* **9**: 1307–1310.

Bowen ID, Bowen SM, and Jones AH. 1998. *Mitosis and Apoptosis*. Chapman & Hall, London.

Brown DA, and London E. 1998. Functions of lipid rafts in biological membranes. *Annual Review of Cell and Developmental Biology* **14**: 111–136.

Callera GE, Montezano ACI, Yogi A, Tostes RCA, and Touyz RM. 2007. Vascular signaling through cholesterol-rich domains: implications in hypertension. *Current Opinion in Nephrology and Hypertension* **16**: 90–104.

Cameron DG, Martin A, Moffatt DJ, and Mantsch HH. 1985. Infrared Spectroscopic Study of the Gel to Liquid-Crystal Phase-Transition in Live Acholeplasma-Laidlawii Cells. *Biochemistry* **24**: 4355–4359.

Chapman D. 1975. Phase-Transitions and Fluidity Characteristics of Lipids and Cell-Membranes. *Quarterly Reviews of Biophysics* **8**: 185–235.

Chintagari NR, Jin N, Wang PC, Narasaraju TA, Chen JW, and Liu L. 2006. Effect of cholesterol depletion on exocytosis of alveolar type II cells. *American Journal of Respiratory Cell and Molecular Biology* **34**: 677–687.

Cordy J, Hooper N, and Turner A. 2006. The involvement of lipid rafts in Alzheimer's disease. *Molecular Membrane Biology* **23**: 111–122.

Damjanovich S, Vereb G, Schaper A, Jenei A, Matko J, Starink JPP, Fox GQ, Arndtjovin DJ, and Jovin TM. 1995. Structural Hierarchy in the Clustering of Hla Class-I Molecules in the Plasma-Membrane of Human Lymphoblastoid-Cells. *Proceedings of the National Academy of Sciences of the United States of America* **92**: 1122–1126.

Dey I, and Farkas T. 1992. Temperature Shifts Induce Adaptive-Changes in the Physical State of Carp (Cyprinus-Carpio L) Erythrocyte Plasma-Membranes Invitro. *Fish Physiology and Biochemistry* **10**: 347–355.

Dietrich C, Yang B, Fujiwara T, Kusumi A, and Jacobson K. 2002. Relationship of lipid rafts to transient confinement zones detected by single particle tracking. *Biophysical Journal* **82**: 274–284.

Durairaj G, and Vijayakumar I. 1984. Temperature-Acclimation and Phospholipid Phase-Transition in Hypothalamic Membrane Phospholipids of Garden Lizard, Calotes-Versicolor. *Biochimica Et Biophysica Acta* **770**: 7–14.

Edidin M. 2003. The state of lipid rafts: From model membranes to cells. *Annual Review of Biophysics and Biomolecular Structure* **32**: 257–283.

Edidin M, Zuniga MC, and Sheetz MP. 1994. Truncation Mutants Define and Locate Cytoplasmic Barriers to Lateral Mobility of Membrane-Glycoproteins. *Proceedings of the National Academy of Sciences of the United States of America* **91**: 3378–3382.

Freeman MR, Cinar B, and Lu ML. 2005. Membrane rafts as potential sites of nongenomic hormonal signaling in prostate cancer. *Trends in Endocrinology and Metabolism* **16**: 273–279.

Hames D, and Hooper N. 2005. *Biochemistry*. Taylor & Francis, New York.

Hammond AT, Heberle FA, Baumgart T, Holowka D, Baird B, and Feigenson GW. 2005. Crosslinking a lipid raft component triggers liquid ordered-liquid disordered phase separation in model plasma membranes. *Proceedings of the National Academy of Sciences of the United States of America* **102**: 6320–6325.

Hessler JA, Budor A, Putchakayala KG, Mecke A, Rieger D, Banaszak Holl MM, Orr BG, Bielinska A, Beals J, and Baker JR. 2005. An atomic force microscopy study of early morphological changes during apoptosis. *Langmuir* **21**: 9280–9286.

Holowka D, Gosse JA, Hammond AT, Han XM, Sengupta P, Smith NL, Wagenknecht-Wiesner A, Wu M, Young RM, and Baird B. 2005. Lipid segregation and IgE receptor signaling: A decade of progress. *Biochimica Et Biophysica Acta-Molecular Cell Research* **1746**: 252–259.

Ishitsuka R, Sato SB, and Kobayashi T. 2005. Imaging lipid rafts. *Journal of Biochemistry* **137**: 249–254.

Jacobson K, Mouritsen OG, and Anderson RGW. 2007. Lipid rafts: at a crossroad between cell biology and physics. *Nature Cell Biology* **9**: 7–14.

Karnell FG, and Monroe JG. 2006. The role of membrane lipids in the regulation of immune cell activity. *Transfusion Medicine and Hemotherapy* **33**: 8–17.

Kay JG, Murray RZ, Pagan JK, and Stow JL. 2006. Cytokine secretion via cholesterol-rich lipid raft-associated SNAREs at the phagocytic cup. *Journal of Biological Chemistry* **281**: 11949–11954.

Klauda JB, Kucerka N, Brooks BR, Pastor RW, and Nagle JF. 2006. Simulation-based methods for interpreting X-ray data from lipid bilayers. *Biophysical Journal* **90**: 2796–2807.

Kusumi A, Nakada C, Ritchie K, Murase K, Suzuki K, Murakoshi H, Kasai RS, Kondo J, and Fujiwara T. 2005. Paradigm shift of the plasma membrane concept from the two-dimensional continuum fluid to the partitioned fluid: High-speed single-molecule tracking of membrane molecules. *Annual Review of Biophysics and Biomolecular Structure* **34**: 351–371.

Kusumi A, Sako Y, and Yamamoto M. 1993. Confined Lateral Diffusion of Membrane-Receptors as Studied by Single-Particle Tracking (Nanovid Microscopy) – Effects of Calcium-Induced Differentiation in Cultured Epithelial-Cells. *Biophysical Journal* **65**: 2021–2040.

Lang F, Ritter M, Gamper N, Huber S, Fillon S, Tanneur V, Lepple-Wienhues A, Szabo I, and Gulbins E. 2000. Cell volume in the regulation of cell proliferation and apoptotic cell death. *Cellular Physiology and Biochemistry* **10**: 417–428.

Lommerse PHM, Spaink HP, and Schmidt T. 2004. In vivo plasma membrane organization: results of biophysical approaches. *Biochimica Et Biophysica Acta-Biomembranes* **1664**: 119–131.

Maeno E, Ishizaki Y, Kanaseki T, Hazama A, and Okada Y. 2000. Normotonic cell shrinkage because of disordered volume regulation is an early prerequisite to apoptosis. *Proceedings of the National Academy of Sciences of the United States of America* **97**: 9487–9492.

Mayor S, and Rao M. 2004. Rafts: Scale-dependent, active lipid organization at the cell surface. *Traffic* **5**: 231–240.

McElhaney RN. 1974a. Effect of alterations in physical state of membrane lipids on ability of Acholeplasma-Laidlawii-B to grow at various temperatures. *Journal of Molecular Biology* **84**: 145–157.

McElhaney RN. 1974b. Effect of membrane-lipid phase transitions on membrane structure and on the growth of Acholeplasma laidlawii B. *Journal of Supramolecular Structure* **2**.

McElhaney RN. 1984. The structure and function of the Acholeplasma Laidlawii plasma membrane. *Biochimica et Biophysica Acta* **779**: 1–42.

McFarland MJ, Terebova EA, and Barker EL. 2006. Detergent-resistant membrane microdomains in the disposition of the lipid signaling molecule anandamide. *American Association of Pharmaceutical Scientists Journal* **8**: E95–E100.

Melchior DL, Morowitz HJ, Sturteva JM, and Tsong TY. 1970. Characterization of Plasma Membrane of Mycoplasma-Laidlawii. 7. Phase Transitions of Membrane Lipids. *Biochimica Et Biophysica Acta* **219**: 114–122.
Miersch S, and Mutus B. 2007. Membrane lipid domains: Techniques for visualization and characterization. *Current Analytical Chemistry* **3**: 81–92.
Mouritsen OG. 1987. Phase transitions in biological membranes. *Annals of the New York Academy of Science* **491**: 166–169.
Mukherjee S, and Maxfield FR. 2004. Membrane domains. *Annual Review of Cell and Developmental Biology* **20**: 839–866.
Munro S. 2003. Lipid rafts: Elusive or illusive? *Cell* **115**: 377–388.
Murphy S, Hiller NL, Harrison T, and Lomasney J. 2006. Lipid rafts and malaria parasite infection of erythrocytes. *Molecular Membrane Biology* **23**: 81–88.
Musch MW, Koomoa DLT, and Goldstein L. 2004. Hypotonicity-induced exocytosis of the skate anion exchanger skAE1 – Role of lipid raft regions. *Journal of Biological Chemistry* **279**: 39447–39453.
Nichols B. 2003. Caveosomes and endocytosis of lipid rafts. *Journal of Cell Science* **116**: 4707–4714.
Okada Y, and Maeno E. 2001. Apoptosis, cell volume regulation and volume-regulatory chloride channels. *Comparative Biochemistry and Physiology a-Molecular and Integrative Physiology* **130**: 377–383.
Parton RG, and Hancock JF. 2004. Lipid rafts and plasma membrane microorganization: insights from Ras. *Trends in Cell Biology* **14**: 141–147.
Parton RG, and Richards AA. 2003. Lipid rafts and caveolae as portals for endocytosis: New insights and common mechanisms. *Traffic* **4**: 724–738.
Pelkmans L, Burli T, Zerial M, and Helenius A. 2004. Caveolin-stabilized membrane domains as multifunctional transport and sorting devices in endocytic membrane traffic. *Cell* **118**: 767–780.
Plowman SJ, and Hancock JF. 2005. Ras signaling from plasma membrane and endomembrane microdomains. *Biochimica Et Biophysica Acta-Molecular Cell Research* **1746**: 274–283.
Pralle A, Keller P, Florin EL, Simons K, and Horber JKH. 2000. Sphingolipid-cholesterol rafts diffuse as small entities in the plasma membrane of mammalian cells. *Journal of Cell Biology* **148**: 997–1007.
Puri N, and Roche PA. 2002. Lipid raft association of SNAP-23 and other SNARE proteins: implications for regulated exocytosis from RBL mast cells. *Molecular Biology of the Cell* **13**: 506A–506A.
Puri N, and Roche PA. 2006. Ternary SNARE complexes are enriched in lipid rafts during mast cell exocytosis. *Traffic* **7**: 1482–1494.
Razin S. 1979. In: Barile MF and Razin S (eds.), *The Mycoplasmas*. New York: Academic Press. 188–211.
Reinert JC, and Steim JM. 1970. Calorimetric Detection of a Membrane-Lipid Phase Transition in Living Cells. *Science* **168**: 1580–1582.
Rejman J, Oberle V, Zuhorn IS, and Hoekstra D. 2004. Size-dependent internalization of particles via the pathways of clathrin-and caveolae-mediated endocytosis. *Biochemical Journal* **377**: 159–169.
Roy R, Das AB, and Ghosh D. 1997. Regulation of membrane lipid bilayer structure during seasonal variation: A study on the brain membranes of Clarias batrachus. *Biochimica Et Biophysica Acta-Biomembranes* **1323**: 65–74.
Rush B. 1785. An inquiry into the effects of ardent spirits upon the human body and mind. *Quaterly Journal of Studies on Alcohol* **4**: 321.
Salaun C, Gould GW, and Chamberlain LH. 2005. Lipid raft association of SNARE proteins regulates exocytosis in PC12 cells. *Journal of Biological Chemistry* **280**: 19449–19453.
Salaun C, James DJ, and Chamberlain LH. 2004. Lipid rafts and the regulation of exocytosis. *Traffic* **5**: 255–264.
Silvius JR, and McElhaney RN. 1978. Growth and membrane lipid properties of Acholeplasma-Laidlawii-B lacking Fatty-Acid heterogeneity. *Nature* **272**: 645–647.

Simons K, and Ikonen E. 1997. Functional rafts in cell membranes. *Nature* **387**: 569–572.
Simons K, and Toomre D. 2000. Lipid rafts and signal transduction. *Nature Reviews Molecular Cell Biology* **1**: 31–39.
Simons K, and Vaz WLC. 2004. Model systems, lipid rafts, and cell membranes. *Annual Review of Biophysics and Biomolecular Structure* **33**: 269–295.
Singer SJ, and Nicolson GL. 1972. Fluid mosaic model of structure of Cell-Membranes. *Science* **175**: 720–731.
Steim JM, Tourtell ME, Reinert JC, McElhaney RN, and Rader RL. 1969. Calorimetric Evidence for Liquid-Crystalline State of Lipids in a Biomembrane. *Proceedings of the National Academy of Sciences of the United States of America* **63**: 104–109.
Stewart GT. 1961. Mesomorphic forms of lipid in structure of normal and Atheromatous Tissues. *Journal of Pathology and Bacteriology* **81**: 385–393.
Stryer L. 1981. *Biochemistry*. W.H. Freeman and Company, New York.
Szabo G, Dolganiuc A, Dai Q, and Pruett SB. 2007. TLR4, ethanol, and lipid rafts: A new mechanism of ethanol action with implications for other receptor-mediated effects. *Journal of Immunology* **178**: 1243–1249.
Taylor D, and Hooper N. 2006. The prion protein and lipid rafts. *Molecular Membrane Biology* **23**: 89–99.
Tepper AD, Ruurs P, Wiedmer T, Sims PJ, Borst J, and van Blitterswijk WJ. 2000. Sphingomyelin hydrolysis to ceramide during the execution phase of apoptosis results from phospholipid scrambling and alters cell-surface morphology. *Journal of Cell Biology* **150**: 155–164.
Vereb G, Matko J, Vamosi G, Ibrahim SM, Magyar E, Varga S, Szollosi J, Jenei A, Gaspar R, Waldmann TA, and Damjanovich S. 2000. Cholesterol-dependent clustering of IL-2R alpha and its colocalization with HLA and CD48 on T lymphoma cells suggest their functional association with lipid rafts. *Proceedings of the National Academy of Sciences of the United States of America* **97**: 6013–6018.
Vereb G, Szollosi J, Matko J, Nagy P, Farkas T, Vigh L, Matyus L, Waldmann TA, and Damjanovich S. 2003. Dynamic, yet structured: The cell membrane three decades after the Singer-Nicolson model. *Proceedings of the National Academy of Sciences of the United States of America* **100**: 8053–8058.
Voet D, and Voet JG. 1995. *Biochemistry*. John Wiley & Sons, New York.
Vu CCQ, Bortner CD, and Cidlowski JA. 2001. Differential involvement of initiator caspases in apoptotic volume decrease and potassium efflux during Fas- and UV-induced cell death. *Journal of Biological Chemistry* **276**: 37602–37611.
Xia FZ, Gao XD, Kwan E, Lam P, Chan LL, Shen L, Gaisano HY, and Tsushima RG. 2004. Lipid raft regulation of beta-cell ion channel gating and exocytosis. *Biophysical Journal* **86**: 170A–170A.
Yang SA, and Endo K. 1994. Acclimation temperature affects activities of 5′-Nucleotidase and Acid-Phosphatase and lipid and Fatty-Acid composition in carp muscle microsomes. *Journal of Food Science* **59**: 1009–1012.
Yang XL, Xiong WC, and Mei L. 2004. Lipid rafts in neuregulin signaling at synapses. *Life Sciences* **75**: 2495–2504.
Yu SP, Canzoniero LMT, and Choi DW. 2001. Ion homeostasis and apoptosis. *Current Opinion in Cell Biology* **13**: 405–411.
Yu SP, and Choi DW. 2000. Ions, cell volume, and apoptosis. *Proceedings of the National Academy of Sciences of the United States of America* **97**: 9360–9362.

Index

A
actin dynamics, 111
action potential, 10–19, 77, 83
acute pancreatitis, 51, 52, 59, 60
airway smooth muscle, 111, 114, 123
alcohol, 51, 60, 86, 146, 147, 160, 177
apoptosis, 63, 68, 70, 174, 175
apoptotic, 63, 69, 70, 174
ATP (Adenosine triphosphate), 111, 118, 119, 127–130, 134, 160, 163, 164, 167
auto/paracrine signalling, 23, 24
autocrine relay, 23, 39
autothixotropy, 153, 155, 156

B
biological membranes, 73–75, 77, 78, 172

C
Ca^{2+}, 1–4, 6, 7, 10–12, 14, 15, 17, 19, 33, 51–61, 75, 78, 80, 82, 164–168
calcium, 12, 19, 52–54, 56, 60, 73, 75, 83, 167
cation-exchange, 1, 2, 7
chain-ordering phase transition, 73
cholesterol, 119, 171, 172, 174–176
collagen, 43, 45–48, 96, 97, 102
contraction mechanism, 159, 163
covalent bonds, 68, 95–106
critical phenomena, 23, 96
crowding (crowded), 48, 49, 112, 134, 135
cytokines, 24
cytoplasm, 48, 49, 54, 65, 68–70,113, 116, 134, 145, 150, 174
cytoskeleton (cytoskeletal), 76, 111–136, 150, 153, 156

D
degeneration, 63
diffusional or nondirectional transport, 24
diffusive messengers, 23
directed percolation, 23, 25, 32, 33, 35–39
divalent-monovalent cation-exchange, 1, 2, 7, 9
dynamics, 28–33, 35, 38, 95, 97, 98, 100, 111–113, 118–124, 128, 130, 132–134, 136, 175

E
enzymes, 52, 59, 67, 95–97, 99, 101, 103, 104, 106, 108, 143, 160
ephemeral polymerisation, 155
ethanol, 14, 15, 51, 52, 54, 56–58, 60, 61, 80, 85, 177
exclusion zones, 143–150
Exocytosis, 73, 77, 80, 81, 83, 86, 88, 89, 172, 176

F
fluid mosaic model, 171–174
function of sleep, 73

G
gel, 1–13, 15, 19, 32, 67–70, 75, 95–108, 146–148, 159–163, 165, 166, 169, 171–177
gelatin gels, 95, 97, 100, 104
general anaesthesia, 73, 79, 87
granules, 51–61, 65

H
helical form, 1
high-pressure neurological syndrome, 73, 86
homeothermy, 73, 88
hydration subcompartments, 43, 49
hydrogen bonding, 44, 45
hydrogen bonds, 1, 44

I
intracellular Ca^{2+}, 51, 52, 59, 168
intracellular perfusion, 10–12, 19
ion exchange, 51–61

J
jamming, 111

L
light- and electron-microscopy, 63, 66, 69
lipid bilayer, 74, 172, 173, 176
lipid raft, 171, 172, 174–176, 177
Lipids, 73, 81, 83, 84, 171, 173, 175

M
magnetic twisting cytometry, 111, 113
MAPK pathway, 24
matrix, 51, 56, 58, 60, 61, 67, 68, 74, 96, 108, 109, 111, 114, 122, 125, 133, 160
messenger molecules, 24–28, 32, 36, 39
microsphere, 117, 146, 147, 149
mode of death, 63
molecular machines, 73–75
monolayer, 44, 46, 48, 76, 83, 144, 173
MPTD machine, 73–75, 77–86, 88–90
myosin, 119

N
Nafion, 147–149
nerve conduction, 18
nerve excitation, 1–20
nerve fiber, structural phase transition in, 1
neuron, dark, 63–66, 68, 69
nonequilibrium phase transition, 23–39

O
oscillation, 37, 89, 155, 167

P
pancreatitis, 51, 52, 59, 60
paracrine cell-to-cell communication, 23
peptide bonds, 96
phase of water, 43, 50
phase transition, 1, 23–39, 43, 44, 63–70, 73–90, 95, 96, 101, 102, 104, 108, 159–169, 171–178
plasma membrane, 49, 60, 68, 81, 83, 119, 143, 171–177
polyanionic gels, 1, 2, 7
polyanionic hydrogels, 1, 10
polymer, 6–8, 60, 74, 95–109, 148, 159–162, 166, 169
pressure-reversal of anaesthesia, 73, 86
protease, 53, 95–97, 100–106, 108
proteolysis (proteolytic), 24, 52, 59, 96, 97, 100, 102, 108

R
random coil, 1, 105, 107
recovery, 7, 8, 63, 68, 73, 89, 125, 127, 128, 130

S
signal transmission, 23, 37, 175
soft glassy rheology, 111–113, 132, 133, 135
soft matter, 95, 131
solid-liquid interface, 145
solvent, 1, 14, 48, 160, 165
spasmoneme, 163–166
spontaneous bead motion, 111, 128–130
stochastic models, 23, 24
structural phase transition, in nerve fiber, 1
swimming, 159, 166–169
synaptic exocytosis, 73, 77, 81, 88, 89
Synechococcus, 159, 160, 163, 166–169

T
tendon, 43, 45–49
thermoregulation, 73, 84, 85, 88, 177
transglutaminases, 95–104, 108
trypsin, 51–61
trypsin activation, 51–61

U
unstirred layers (USL), 143–159

V
volume phase transition, 159–169
vorticellid, 159, 160, 163, 166, 168, 169

W
water bridge, 44–46, 48

Z
zymogen granules, 51–61

MIX
Papier aus verantwortungsvollen Quellen
Paper from responsible sources
FSC® C105338

Printed by Books on Demand, Germany